Ergodic Theory of Numbers

© 2002 by
The Mathematical Association of America (Incorporated)
Library of Congress Catalog Card Number 2002101376

Complete Set ISBN 0-88385-000-1
Vol. 29 ISBN 0-88385-034-6

Printed in the United States of America

Current Printing (last digit):
10 9 8 7 6 5 4 3 2 1

The Carus Mathematical Monographs

Number Twenty-Nine

Ergodic Theory of Numbers

Karma Dajani
University of Utrecht

Cor Kraaikamp
Delft University of Technology

Published and Distributed by
THE MATHEMATICAL ASSOCIATION OF AMERICA

THE
CARUS MATHEMATICAL MONOGRAPHS

Published by
THE MATHEMATICAL ASSOCIATION OF AMERICA

The following Monographs have been published:

MAA Service Center
P. O. Box 91112
Washington, DC 20090-1112
800-331-1MAA   FAX: 301-206-9789

To Rafael

Contents

Preface

In this book we will look at the interaction between two fields of mathematics: number theory and ergodic theory (as part of dynamical systems). The subject under study is thus part of what is known in France as *Théorie Ergodique des Nombres*, and consists of a family of series expansions of numbers in the unit interval [0, 1] with their 'metrical properties.' So the questions we want to study are number theoretical in nature, and the answers will be obtained with the help of ergodic theory. That is, we will view these expansions as iterations of an appropriate measure-preserving transformation on [0,1], which will then be shown to be ergodic. The number-theoretical questions will be reformulated in the language of ergodic theory. What it means to be ergodic, or—in general—what the basic ideas behind ergodic theory entail, will be explained along the way.

This book grew out of a course given in 1996 at George Washington University, Washington, DC, during the Summer Program for Women in Mathematics, sponsored by NSA. Our aim was not to write yet another book on ergodic theory (there are already several outstanding books, most of them mentioned in these pages), but to introduce first-year graduate students to a dynamical way of thinking. Consequently, many classical concepts from ergodic theory are either briefly mentioned, or even left out. In this book we focus our attention on easy concepts like ergodicity and the ergodic theorem, and then apply these concepts to familiar expansions to obtain old and new results in an elegant and straightforward manner.

Clearly this means that a number of concepts from probability and measure theory will be used. In our set-up we first introduce these, in—we hope and think—an informal and gentle way.

We thank the directors Murli M. Gupta, Robbie Robinson and Dan Ullman of the Summer Program for Women in Mathematics, and the participants Alissa Andreichuk, Christine Collier, Amy Cottrell, Lisa Darlington, Christy Dorman, Julie Frohlich, Molly Kovaka, Renee Yong, Ran Liu, Susan Matthews, Gail Persons, Jakayla Robbins, Beth Samuels, Elizabeth Trageser, Sharon Tyree, Meta Voelker and Joyce Williams of this summer program, who were faced with a preliminary version of this book. Their remarks, comments, improvements and enthusiasm helped us tremendously to improve the original notes.

We also thank Ken Ross and Harold Boas of the MAA, whose constructive criticism, sharp observations and patience changed the original manuscript into a readable text.

Karma Dajani
Utrecht, The Netherlands

Cor Kraaikamp
Delft, The Netherlands

CHAPTER 1
Introduction

Let x be a real number from the unit interval $[0,1)$. As is well known, x can be written as

$$x = \sum_{k=1}^{\infty} \frac{a_k(x)}{10^k} \, , \qquad (1.1)$$

where $a_k = a_k(x) \in \{0, 1, \dots, 9\}$ for $k \geq 1$. We will denote (1.1) by $x = .a_1 a_2 \dots a_k \dots$. This expansion of x, the so-called *decimal expansion*, is unique if we do not allow infinite expansions with $a_k = 9$ from some k_0 on, or conversely, if we allow only infinite expansions. So for example we can write the (rational) number $x = 10^{-2}$ in the following two ways, as either $.01$ or $.0099999\dots99999\dots$.

Decimal expansions have become such an integral part of our daily life that it takes some thought to realize that these expansions are mysterious objects. One quickly stumbles upon easy questions like: *Why is the decimal expansion of $\frac{1}{3}$ infinite, while the expansion of $\frac{1}{4}$ is finite?* and *Why is a number rational if and only if its decimal expansion is eventually-periodic?* Harder questions are: *In an arbitrary number why are 10% of the digits equal to 7?* and *What percentage of the digits of π are equal to 7?* A third hard question could be: *Is there another/better way to represent numbers?*

In this book we will look at the first and third hard questions (this does not imply, however, that we will ignore easy questions; in fact, often easy questions turn out to be pretty hard!). Essential in dealing

1

with these questions will be (the interplay of) techniques from number theory and ergodic theory. It should be noted that the second hard question is beyond the current state of affairs in mathematics; even if we replace π by $\sqrt{2}$ nothing is known!

1.1 Decimal expansions of rational numbers

In this section we will see that the decimal expansion of a rational number x can easily be characterized: it is either finite or eventually-periodic. Conversely, if the decimal expansion of x is finite or eventually-periodic, then x is a rational number.

Some examples of finite expansions are

$$\frac{13}{20} = .65 \text{ and } \frac{171951}{31250000} = .005502432 \, ,$$

while

$$\frac{1}{3} = .333\cdots = .\overline{3} \text{ and } \frac{1}{13} = .076923076923\cdots = .\overline{076923}$$

are examples of purely-periodic expansions (the bar indicates the period); see below for a formal definition. Examples of eventually-periodic expansions are

$$\frac{1}{6} = .1\overline{6} \text{ and } \frac{5}{14} = .3\overline{571428} \, .$$

Exercise 1.1.1. Recall how these expansions can be obtained, and try a few for yourself. Do you see a pattern?   ∎

Now assume that $x \in [0, 1)$, with finite decimal expansion $x = .a_1 a_2 \cdots a_k$, with $a_i \in \{0, 1, \ldots, 9\}$ and $a_k \neq 0$. Clearly

$$x = \frac{a_1 \cdot 10^{k-1} + a_2 \cdot 10^{k-2} + \cdots + a_{k-1} \cdot 10 + a_k}{10^k} \in \mathbb{Q} \, ,$$

where $\mathbb{Q}$ is the set of rationals.

Next suppose x has an eventually-periodic decimal expansion

$$x = .a_1 \cdots a_\ell \overline{a_{\ell+1} \cdots a_{\ell+n}} \; ;$$

in case both ℓ and n are chosen minimal we call $a_1 \cdots a_\ell$ the *pre-period* of x and $a_{\ell+1} \cdots a_{\ell+n}$ the *period* of x. In case $\ell = 0$ we call x *purely-periodic*, otherwise x is called *eventually-periodic*.

Setting

$$y = .a_1 \cdots a_\ell \text{ and } z = .a_{\ell+1} \cdots a_{\ell+n} \,,$$

one has that $y, z \in \mathbb{Q}$.

Exercise 1.1.2. Using the above notations, and setting $\omega = \overline{.a_{\ell+1} \cdots a_{\ell+n}}$, show that

$$\omega = \frac{10^n}{10^n - 1} z.$$

Use this to find that

$$x = y + \frac{10^{n-\ell}}{10^n - 1} z. \qquad \blacksquare$$

Now suppose that $x = p/q$, with $p \in \mathbb{Z}$ and $q \in \mathbb{N}$. We moreover assume (just for convenience) that $x \in (0, 1)$ and that $(p, q) = 1$, i.e., p and q are relatively prime. From our first two examples one sees that, if $q = 2^k 5^m$, where k and m are nonnegative integers, one has that x has a finite decimal expansion, i.e.,

$$\frac{13}{20} = \frac{13}{2^2 \cdot 5} = \frac{13 \cdot 5}{2^2 \cdot 5^2} = \frac{65}{10^2} = .65$$

and

$$\frac{171951}{31250000} = \frac{171951}{2^4 \cdot 5^9} = \frac{171951 \cdot 2^5}{10^9}$$

$$= \frac{5502432}{10^9} = .005502432 \,.$$

Exercise 1.1.3. Let $x = p/q$, with $p \in \mathbb{Z}$, $q \in \mathbb{N}$ and $(p, q) = 1$. Show that x has a finite decimal expansion if and only if $q = 2^k 5^m$, where k and m are non-negative integers.   ∎

Let us have a closer look at our other examples. The expansions of fractions like $\frac{5}{14}$ and $\frac{1}{6}$ were obtained simply by applying a *division algorithm* from our pre-high school days, which we now discuss. We will see that the remainder terms (not the digits in the expansion) play a key role.

Obviously each remainder term has only q possibilities. Now if we denote the remainder term after the ith division by r_i, where $r_0 = p$, then there exist positive integers k and m such that $r_k = r_{k+m}$, from which it follows that

$$r_{k+1} = r_{k+m+1}, \; r_{k+2} = r_{k+m+2}, \dots.$$

But then one has that

$$a_{k+1} = a_{k+m+1}, \; a_{k+2} = a_{k+m+2}, \dots, \; a_{k+m+1} = a_{k+2m+1};$$

since $a_{k+1} = a_{k+m+1}$ it follows that the expansion is periodic.

For example, if we apply the division algorithm to $\frac{5}{14}$, we find that

$$r_0 = 5, \; r_1 = 8, \; r_2 = 10, \; r_3 = 2, \; r_4 = 6, \; r_5 = 4, \; r_6 = 12,$$

and

$$a_1 = 3, \; a_2 = 5, \; a_3 = 7, \; a_4 = 1, \; a_5 = 4, \; a_6 = 2, \; a_7 = 8.$$

The seventh remainder term r_7 is again 8, and from now on the pattern starts to repeat itself. We see that $k = 2$ and $m = 6$.

The last four examples seem to suggest that $x \in \mathbb{Q} \cap [0, 1)$ is purely-periodic if and only if $(q, 10) = 1$. For example, note that $(q, 10) = (14, 10) = 2 \neq 1$; indeed, the expansion of $\frac{5}{14}$ is not purely-periodic.

Now suppose that $(q, 10) = 1$, and let k and m be as before, i.e., $r_k = r_{k+m}$. By the division algorithm one has

$$10 \cdot r_{k-1} = a_k \cdot q + r_k \quad \text{and} \quad 10 \cdot r_{k+m-1} = a_{k+m} \cdot q + r_{k+m}$$

and therefore

$$10(r_{k-1} - r_{k+m-1}) = q(a_k - a_{k+m}). \qquad (1.2)$$

Exercise 1.1.4. Show that (1.2) implies that $r_{k-1} = r_{k+m-1}$. (*Hint:* Recall that $0 \le r_i < q$ or that $0 \le a_i < 10$). ∎

Repeating the above argument yields that $r_0 = r_m$, i.e., x has a purely-periodic decimal expansion. Conversely, if the expansion of p/q is purely-periodic, then all $r_i > 0$. In particular we have $0 < r_1 < q$, and since $r_0 = p$, $10 \cdot r_0 = a_1 q + r_1$, it follows from the assumption $(p, q) = 1$ that $(q, 10) = 1$.

1.2 Another look at the decimal expansion

In this section we will show how the decimal digits a_n can be obtained in a dynamical way. This leads us in a natural way to measure theory. We then take the opportunity to review some basic concepts and results from measure theory which will be used throughout this book. Also some terminology from ergodic theory will be discussed.

1.2.1 How are the digits a_n obtained?

The idea is to make a partition of $[0,1)$ into intervals of the form $[\frac{i}{10}, \frac{i+1}{10})$, where $i = 0, 1, \ldots, 9$. Label the interval $[\frac{i}{10}, \frac{i+1}{10})$ with the digit i and write $a_1(x) = i$ for any $x \in [\frac{i}{10}, \frac{i+1}{10})$. To find a_2 we partition each of the 10 intervals $[\frac{i}{10}, \frac{i+1}{10})$ into 10 pieces of equal length. This yields intervals $[\frac{i}{10} + \frac{j}{10^2}, \frac{i}{10} + \frac{j+1}{10^2})$, $0 \le i \le 9$ and $0 \le j \le 9$. For $x \in [\frac{i}{10} + \frac{j}{10^2}, \frac{i}{10} + \frac{j+1}{10^2})$ we write $a_1(x) = i$, $a_2(x) = j$. To determine the third digit, you must subdivide each of $[\frac{i}{10} + \frac{j}{10^2}, \frac{i}{10} + \frac{j+1}{10^2})$ into 10 pieces of equal length, etc.

So, the digits are determined by repeated application of the same operation, which is subdividing intervals into 10 pieces of the same length. Let us have another look at this mechanism. Consider the map $T : [0, 1) \rightarrow [0, 1)$ given by

$$Tx = 10x \,(\text{mod } 1) = \begin{cases} 10x, & 0 \le x < \dfrac{1}{10}, \\ 10x - 1, & \dfrac{1}{10} \le x < \dfrac{2}{10}, \\ \vdots & \vdots \\ 10x - 9, & \dfrac{9}{10} \le x < 1. \end{cases} \quad (1.3)$$

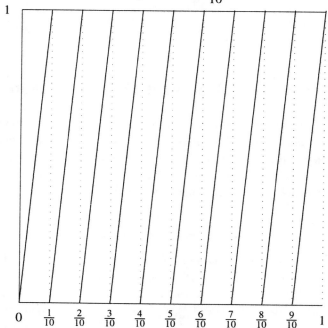

Figure 1.1. The decimal map T

In other words, T is given by

$$Tx = 10x - i \quad \text{if} \quad \frac{i}{10} \le x < \frac{i+1}{10}, \quad i = 0, 1, \dots, 9.$$

This process is illustrated in Figure 1.1.

Thus, if $x = \sum_{k=1}^{\infty} \frac{a_k}{10^k} = .a_1 a_2 \cdots$, then $Tx = \sum_{k=1}^{\infty} \frac{a_{k+1}}{10^k} = .a_2 \cdots a_k \cdots$, so that $a_1(Tx) = a_2(x)$ and in general $a_1(T^n x) = a_{n+1}(x)$, where $T^n x$ denotes the n-fold application of T to x. Notice that we mixed here two ways of representing x: as an infinite series, and as an infinite sequence of symbols. See also Exercise 1.2.16 and the remarks preceding it.

If we denote the greatest integer not exceeding ξ by $\lfloor \xi \rfloor$, then we clearly have that $Tx = 10x - \lfloor 10x \rfloor = 10x - a_1(x)$, from which it follows that

$$x = \frac{a_1(x)}{10} + \frac{Tx}{10}$$

$$= \frac{a_1(x)}{10} + \frac{a_2(x)}{10^2} + \frac{T^2 x}{10^2}$$

$$\vdots$$

$$= \frac{a_1(x)}{10} + \cdots + \frac{a_k(x)}{10^k} + \frac{T^k x}{10^k}.$$

Notice that if $x = \ell/10^k$, then $T^k x = 0$, and

$$x = \frac{a_1(x)}{10} + \cdots + \frac{a_k(x)}{10^k}.$$

If $T^k x \ne 0$ for any $k \ge 0$, taking limits gives

$$x = \frac{a_1(x)}{10} + \cdots + \frac{a_k(x)}{10^k} + \cdots.$$

Thus we see that under iteration of T no infinite string of 9's can occur in the expansion of any point. In this respect, all points in $[0, 1)$ have unique expansions.

Intuitively, it is clear that for a generic $x \in [0, 1)$ one must have that $a_1(x)$ attains any of the values from $\{0, 1, \ldots, 9\}$ with the same probability, i.e., $\frac{1}{10}$.

1.2.2 Measure theory, isomorphisms and shift spaces

Since the concept of probability measure is one of the basic concepts of this book, we will give a heuristic introduction to it. Let Ω be the set of all possible outcomes of an experiment, and let A be any subset of Ω (such a subset is called an *event*). Assume for the moment that Ω is at most countable, e.g., $\Omega = \{1, 2, \ldots, 6\}$ and $A = \{1\}$; then we define $P(A)$, the *probability of A*, as

$$\lim_{n \to \infty} \frac{N(n, A)}{n}, \qquad (1.4)$$

where $N(n, A)$ is the number of times the event A occurs in n independently performed identical experiments. Clearly one should show that the limit in (1.4) exists, which follows from the *Law of Large Numbers* from probability theory; see also Chapter 3, [Bil95] and [Fel71]. Let 2^Ω be the collection of all subsets of Ω, also known as *the power set* of Ω. Now P is a set-function from the power set 2^Ω to the interval $[0, 1]$, satisfying

(i) $P(\Omega) = 1$;

(ii) if $A \cap B = \emptyset$, then $P(A \cup B) = P(A) + P(B)$.

Exercise 1.2.1. Show that (i) and (ii) follow at once from our heuristic approach. ∎

Exercise 1.2.2. In the above heuristic approach we chose $\Omega = \{1, 2, \ldots, 6\}$ and $A = \{1\}$. Which classical example did we have in mind? If no data are available, how would you choose $P(A)$ in that case? ∎

In general one cannot use the above heuristic approach. Obviously it is impossible to do infinitely many experiments to determine $P(A)$, so $P(A)$ has to be chosen, possibly using available data, or from assumptions underlying your experiment. For example, let $\Omega = [0, 1]$, and let $0 \le a < b \le 1$. Then the probability that an arbitrary $x \in [0, 1]$ lies in the interval $[a, b]$ equals

$$\lambda([a, b]) := b - a = \text{ the length of } [a, b], \qquad (1.5)$$

and one has that λ satisfies (i) and (ii) for any two intervals A and B in $[0, 1]$. In fact λ also satisfies

(ii)* If $A_1, A_2, \ldots$ are intervals or complements of intervals in $[0, 1]$ such that $A_k \cap A_\ell = \emptyset$ whenever $k \neq \ell$, then

$$\lambda\left(\bigcup_{n=1}^{\infty} A_n\right) = \sum_{n=1}^{\infty} \lambda(A_n).$$

Exercise 1.2.3. Let $A = [0, 1] \cap \mathbb{Q}$. Show that $\lambda(A) = 0$. ∎

Exercise 1.2.4. Let $C_1 := [0, 1]$, $C_2 := [0, 1] \setminus (\frac{1}{3}, \frac{2}{3})$, and let $C_3 := C_2 \setminus ((\frac{1}{9}, \frac{2}{9}) \cup (\frac{7}{9}, \frac{8}{9}))$, i.e., C_2 is obtained from C_1 by removing the middle third interval of C_1, and C_3 is obtained from C_2 by removing the middle third interval from each of its two intervals. In general C_{n+1} is obtained from C_n by removing from each of the 2^{n-1} intervals of C_n the middle third interval.

(a) Determine for each $n \ge 1$ the length $\lambda(C_n)$ of C_n.
(b) Let $C_\infty = \bigcap_{n=1}^{\infty} C_n$; show that $\lambda(C_\infty) = 0$. ∎

In general, if X is a set, a family of subsets $\mathcal{F}$ of X is said to be a *σ-algebra* if the following conditions hold

(i) $X \in \mathcal{F}$;
(ii) if $A \in \mathcal{F}$, then $A^c \in \mathcal{F}$;
(iii) if $A_1, A_2, \ldots \in \mathcal{F}$, then $A = \bigcup_{n=1}^{\infty} A_n \in \mathcal{F}$.

The pair $(X, \mathcal{F})$ is referred to as a *measure space*.

Exercise 1.2.5. Let $\mathcal{F}_j$ for $j \in J$ be a family of σ-algebras on X, and let $\mathcal{F} = \bigcap_{j \in J} \mathcal{F}_j$. Show that $\mathcal{F}$ is a σ-algebra on X. ∎

Now let $\mathcal{A}$ be some collection of subsets of X. We say that the σ-algebra $\mathcal{F}$ is *generated by* $\mathcal{A}$ if $\mathcal{F}$ is the smallest σ-algebra containing $\mathcal{A}$, and we write $\mathcal{F} = \sigma(\mathcal{A})$. In particular, if $X = [0, 1]$, and $\mathcal{A}$ is the collection of intervals (a, b) in $[0, 1]$, then the so-called Borel σ-algebra $\mathcal{B}$ is the σ-algebra generated by $\mathcal{A}$. An element of $\mathcal{B}$ is called a Borel set. We call $\mathcal{A}$ a *semi-algebra* if (i) $\mathcal{A}$ is closed under finite intersections and (ii) the complement of any set in $\mathcal{A}$ is a finite disjoint union of elements in $\mathcal{A}$.

Exercise 1.2.6. Consider $[0, 1)$ with the Borel σ-algebra $\mathcal{B}$. Show that the family $\mathcal{A}$ of all half open−half closed intervals $[a, b)$ in $[0, 1)$ is a generating semi-algebra for $\mathcal{B}$. ∎

In the language of measure theory, the concept of probability as introduced above is called a (normalized) measure. In general, a *measure* on $(X, \mathcal{F})$ is a set-function $\mu : \mathcal{F} \to [0, \infty)$ satisfying $\mu(\emptyset) = 0$ and

$$\mu\left(\bigcup_{n=1}^{\infty} A_n\right) = \sum_{n=1}^{\infty} \mu(A_n),$$

whenever $A_1, A_2, \ldots$ is a pairwise disjoint collection from $\mathcal{F}$. We call the triplet $(X, \mathcal{F}, \mu)$ a *finite measure space*. We call μ a *probability measure* if $\mu(X) = 1$. In this book all measures under consideration will be probability measures. A finite measure space $(X, \mathcal{F}, \mu)$ is *complete* if for every $B \in \mathcal{F}$ with $\mu(B) = 0$ one has $C \in \mathcal{F}$ for every $C \subset B$.

One can show that λ from (1.5) can be extended to a measure on the Borel σ-algebra $\mathcal{B}$ on $[0, 1]$, using the well-known Carathéodory Theorem on extending a measure on a generating semi-algebra to the whole σ-algebra; see [Roy88]. If we also denote this measure by λ then we get what is usually known as the *Borel measure* (on $[0, 1]$). However, the Borel σ-algebra is not complete. One can show that the cardinality of $\mathcal{B}$ is that of the continuum, while the collection of all subsets of the Cantor set C_∞ has cardinality equal to the cardinality of the power set of the continuum. But then there exists a subset of the Cantor set that is not Borel measurable, and hence $\mathcal{B}$ is not complete, see [KT66] and [Hal50].

One can extend λ on a complete σ-algebra containing $\mathcal{B}$, known as the *Lebesgue σ-algebra* $\mathcal{L}$. In fact, $\mathcal{L}$ consists of all sets of the form $B \cup N$, where $B \in \mathcal{B}$ and N is a subset of a set in $\mathcal{B}$ with Borel measure zero. We extend λ to elements of $\mathcal{L}$ as follows:

$$\lambda^*(B \cup N) = \lambda(B).$$

To simplify notation, we denote λ^* again by λ, and we call λ the *Lebesgue measure* on $[0, 1]$. In the same way any probability space $(X, \mathcal{F}, \mu)$ can be extended to a complete space $(X, \mathcal{C}, \mu^*)$; as above we also denote μ^* by μ. The σ-algebra $\mathcal{C}$ is called the completion of $\mathcal{F}$ under μ.

We remark now that every subset of the Cantor set is Lebesgue measurable. As a result of this one sees that $\mathcal{L}$ and the power set of the continuum have the same cardinality. However, one can construct non Lebesgue measurable subsets of $[0, 1]$, see again [KT66] and [Hal50].

The concept of Borel and Lebesgue σ-algebra can be extended in a natural way to $[0, 1]^2$. One then speaks of the *product Borel σ-algebra* $\mathcal{B} \times \mathcal{B}$ generated by the open (or semi-open, or closed) rectangles of the form $(a, b) \times (c, d)$ (or the appropriate modification). The corresponding product Borel measure $\lambda \times \lambda$ is defined on rectangles $(a, b) \times (c, d)$ by

$$(\lambda \times \lambda)\,[(a, b) \times (c, d)] = (b - a)(d - c).$$

The completion of $\mathcal{B} \times \mathcal{B}$ under $\lambda \times \lambda$ is the *product Lebesgue σ-algebra* $\mathcal{L} \times \mathcal{L}$. The extension of $\lambda \times \lambda$ on $\mathcal{L} \times \mathcal{L}$ is also denoted by $\lambda \times \lambda$, and is called product Lebesgue measure.

The following approximation theorem plays an important role in this book; for a proof, see [KT66], p. 84.

Theorem 1.2.7. *Let* $(X, \mathcal{F}, \mu)$ *be a probability space, and let* $\mathcal{A}$ *be a semi-algebra such that* $\mathcal{F} = \sigma(\mathcal{A})$. *Furthermore, let* $\mathcal{C}$ *be the collection of all sets that are finite disjoint unions of elements from* $\mathcal{A}$. *Then for every* $B \in \mathcal{F} = \sigma(\mathcal{A})$ *and for every* $\varepsilon > 0$ *there exists* $C \in \mathcal{C}$ *such that* $\mu(B \triangle C) < \varepsilon$. *Here* $B \triangle C$ *denotes the symmetric difference of* B *and* C, *defined by* $(B \cup C) \setminus (B \cap C)$.

Let $(X, \mathcal{F})$ and $(Y, \mathcal{G})$ be measure spaces. A function $f : X \rightarrow Y$ is said to be *measurable* if

$$f^{-1}(B) := \{x \in X; \ f(x) \in B\} \in \mathcal{F},$$

for any $B \in \mathcal{G}$. In particular, if $(Y, \mathcal{G})$ is the the real line with the Borel σ-algebra $\mathcal{B}$, then f is called Borel measurable, or $\mathcal{B}$-measurable.

Remark 1.2.8. It is a well-known theorem in measure theory that $f : X \rightarrow Y$ is measurable if and only if $f^{-1}(B) \in \mathcal{F}$ for every $B \in \mathcal{A}$, where $\mathcal{A}$ is a generating semi-algebra of $\mathcal{G}$; see e.g. [Rud87]. In particular, $f : X \rightarrow \mathbb{R}$ is $\mathcal{B}$-measurable if $f^{-1}(a, b) \in \mathcal{F}$ for every interval (a, b). ∎

Exercise 1.2.9. Let $(X, \mathcal{F})$ and $(Y, \mathcal{G})$ be measure spaces and let $f : X \rightarrow Y$ be a measurable function. If μ is a measure on $(X, \mathcal{F})$, show that ν defined by $\nu(C) := \mu\left(f^{-1}(C)\right)$, where $C \in \mathcal{G}$, is a measure on $(Y, \mathcal{G})$. We denote ν by $f * \mu$ and we call it the pull-back or lifted measure of μ. ∎

Exercise 1.2.10. Let $(X, \mathcal{F})$ be a measure space, and $f_n : X \rightarrow \mathbb{R}$ be a sequence of $\mathcal{B}$-measurable functions. Define g on X by

$$\bar{f}(x) = \limsup_{n \to \infty} f_n(x).$$

Show that $\bar{f}$ is $\mathcal{B}$-measurable. Show that the same holds for $\underline{f} = \liminf_{n \to \infty} f_n$. ∎

Exercise 1.2.11. Let $(X, \mathcal{O})$ be a topological space, where $\mathcal{O}$ is the collection of all open sets of X, and let $\mathcal{F} = \sigma(\mathcal{O})$. Then $\mathcal{F}$ is again called a *Borel σ-algebra*. Show that every function $f : X \to \mathbb{R}$ that is $\mathcal{O}$-continuous is also $\mathcal{F}$-measurable. ∎

In case $(X, \mathcal{F}, P)$ is a probability space (i.e., P is a probability measure on $(X, \mathcal{F})$), we usually write Ω instead of X, and measurable functions are usually called *random variables* and denoted by capital letters X, Y, Z, ...

Let us look at $[0,1)$ as a measure space equipped with the Lebesgue σ-algebra $\mathcal{L}$. Let λ be the Lebesgue measure on $([0, 1), \mathcal{L})$, so the Lebesgue measure $\lambda[a, b)$ of any interval $[a, b)$ (or (a, b), $(a, b]$, etc.) is just the length of the interval, namely $b - a$. We once more consider the decimal map T. Now let $[a, b) \subset [0, 1)$; we know that $\lambda[a, b) = b - a$. What is $\lambda(T^{-1}[a, b)) = \lambda(\{x : Tx \in [a, b)\})$? (See also Figure 1.1.)

To answer this question, consider

$$\lambda\big(T^{-1}[a, b)\big) = \lambda\left(\bigcup_{k=0}^{9}\left[\frac{k}{10} + \frac{a}{10}, \frac{k}{10} + \frac{b}{10}\right)\right)$$

$$= \sum_{k=0}^{9}\lambda\left[\frac{k}{10} + \frac{a}{10}, \frac{k}{10} + \frac{b}{10}\right)$$

$$= b - a = \lambda[a, b).$$

In fact T is measure preserving, which is defined as follows.

Definition 1.2.12. *Let $(X, \mathcal{F}, \mu)$ be a probability space. A measurable transformation $T : X \to X$ is measure preserving with respect to μ (equivalently: μ is T-invariant, or μ is an invariant measure for T), if $\mu(T^{-1}A) = \mu(A)$ for all $A \in \mathcal{F}$.*

Remark 1.2.13. In case T is invertible the above definition is equivalent to $\mu(TA) = \mu(A)$ for all $A \in \mathcal{F}$. ∎

The notions of measurability and measure-preservingness of a transformation T on a probability space $(X, \mathcal{F}, \mu)$ are preserved when one passes to the completion $\mathcal{C}$ of $\mathcal{F}$ under μ.

Lemma 1.2.14. *Let $(X, \mathcal{F}, \mu)$ be a probability space, and let the σ-algebra $\mathcal{C}$ be the completion of $\mathcal{F}$ under μ. Then any transformation $T : X \to X$ that is measurable and measure preserving on $(X, \mathcal{F}, \mu)$ has the same property on $(X, \mathcal{C}, \mu)$.*

Proof. Let $C \in \mathcal{C}$. We need to show that $T^{-1}(C) \in \mathcal{C}$ and $\mu(T^{-1}(C)) = \mu(C)$.

Since $\mathcal{C}$ is the completion of $\mathcal{F}$, C has the form $C = B \cup D$, with $B \in \mathcal{F}$, and D is a subset of a set $N \in \mathcal{F}$ of μ-measure zero. Since T is measurable and measure preserving on $(X, \mathcal{F}, \mu)$, it follows that $T^{-1}(B)$, $T^{-1}(N) \in \mathcal{F}$, and $\mu(T^{-1}(N)) = 0$. Thus $T^{-1}(C) = T^{-1}(B) \cup T^{-1}(D) \in \mathcal{C}$.

By definition of the (extended) measure μ on $\mathcal{C}$, one has $\mu(C) = \mu(B)$, and $\mu(T^{-1}(C)) = \mu(T^{-1}(B)) = \mu(B)$, thus $\mu(C) = \mu(T^{-1}(C))$. ∎

Remark 1.2.15. Using Theorem 1.2.7 one can show that any map T on a probability space $(X, \mathcal{F}, \mu)$ is measurable and measure preserving if $\mu\left(T^{-1}(A)\right) = \mu(A)$ for any A in a semi-algebra $\mathcal{A}$ generating $\mathcal{F}$. By Lemma 1.2.14, the same holds true if we replace $\mathcal{F}$ by its completion under μ. In particular any map $T : [0, 1) \to [0, 1)$ is measurable and measure preserving on $(X, \mathcal{L}, \mu)$ if $\mu(T^{-1}A) = \mu(A)$ for every interval $A \subset [0, 1)$; see [Wal82], Theorem 1.1. The above is true if intervals are replaced by elements of a generating semi-algebra. For instance, the decimal map T is a measure preserving transformation with respect to Lebesgue measure. ∎

Let us now consider the so-called *cylinder sets* (this is the terminology used by ergodic theorists) of rank (or order) n, also known (by number theorists) as *fundamental intervals* $\Delta_n = \Delta(i_1, i_2, \ldots, i_n)$ of rank n, defined by

$$\Delta(i_1, i_2, \ldots, i_n)$$
$$:= \{x \in [0, 1) : a_1(x) = i_1, a_2(x) = i_2, \ldots, a_n(x) = i_n\},$$

where $0 \le i_j \le 9$ for each $1 \le j \le n$.

Exercise 1.2.16. Show that the above defined cylinders Δ_n form a semi-algebra generating the Borel σ-algebra $\mathcal{B}$. ∎

For the decimal map T, it is very easy to describe explicitly these intervals Δ_n and obtain their Lebesgue measure.

Order 1:

$$\lambda(\Delta(i)) = \lambda(\{x : a_1(x) = i\}) = \lambda\left(\left[\frac{i}{10}, \frac{i+1}{10}\right)\right) = \frac{1}{10},$$

so the first digit $a_1 = a_1(x)$ is a random variable that is distributed according to the discrete uniform distribution on $\{0, 1, \ldots, 9\}$.

Order 2:

$$\lambda(\Delta(i, j)) = \lambda(\{x : a_1(x) = i, a_2(x) = j\})$$
$$= \lambda\left(\left[\frac{i}{10} + \frac{j}{10^2}, \frac{i}{10} + \frac{j+1}{10^2}\right)\right)$$
$$= \frac{1}{10^2} = \frac{1}{100} = \lambda(\Delta(i))\lambda(\Delta(j)).$$

In general, we have

Order n:

$$\lambda(\Delta(i_1, i_2, \ldots, i_n))$$
$$= \lambda\left(\left[\frac{i_1}{10} + \frac{i_2}{10^2} + \cdots + \frac{i_n}{10^n}, \frac{i_1}{10} + \cdots + \frac{i_n + 1}{10^n}\right)\right)$$

$$= \frac{1}{10^n} = \lambda(\Delta(i_1))\lambda(\Delta(i_2)) \cdots \lambda(\Delta(i_n)).$$

From the above it is now easy to conclude that the digit functions $a_1(x)$, $a_2(x)$, ..., which are random variables on $[0, 1)$, are independent identically distributed (we abbreviate this by i.i.d.) with the same discrete uniform distribution on $\{0, 1, \ldots, 9\}$.

Definition 1.2.17. *A dynamical system is a quadruple* $(X, \mathcal{F}, \rho, T)$, *where X is a non-empty set, $\mathcal{F}$ is a σ-algebra on X, ρ is a probability measure on $(X, \mathcal{F})$ and $T : X \to X$ is a surjective ρ-measure preserving transformation. Further, if T is injective, then we call $(X, \mathcal{F}, \rho, T)$ an invertible dynamical system.*

If $(X, \mathcal{F}, \rho, T)$ is a dynamical system, and $x \in X$, we call the sequence

$$x, Tx, \ldots, T^n x, \ldots$$

the *T-orbit* of x. In case T is invertible, the two-sided T-orbit of x is

$$\ldots, T^{-2}x, T^{-1}x, x, Tx, T^2 x, \ldots.$$

Given two dynamical systems $(X, \mathcal{F}, \rho, T)$ and $(Y, \mathcal{C}, \nu, S)$, what should we mean by: *these systems are the same*? On each space there are two important structures:

(1) The measure structure given by the σ-algebra and the probability measure. Note, that in this context, sets of measure zero can be ignored.

(2) The dynamical structure, given by a measure preserving transformation.

So our notion of *being the same* must mean that we have a map

$$\psi : (X, \mathcal{F}, \rho, T) \to (Y, \mathcal{C}, \nu, S)$$

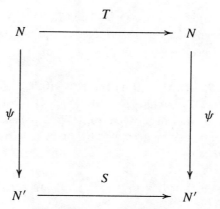

Figure 1.2. ψ and T commute

satisfying

(i) ψ is one-to-one and onto a.e. By this we mean, that if we remove a (suitable) set N_X of measure 0 in X, and a (suitable) set N_Y of measure 0 in Y, the map $\psi : X \setminus N_X \to Y \setminus N_Y$ is a bijection.

(ii) ψ is measurable, i.e., $\psi^{-1}(C) \in \mathcal{F}$, for all $C \in \mathcal{C}$.

(iii) ψ preserves the measures: $\nu = \rho \circ \psi^{-1}$, i.e., $\nu(C) = \rho\left(\psi^{-1}(C)\right)$ for all $C \in \mathcal{C}$.

Finally, we should have that

(iv) ψ preserves the dynamics of T and S, i.e., $\psi \circ T = S \circ \psi$, which is the same as saying that the diagram in Figure 1.2 commutes.

This means that T-orbits are mapped to S-orbits:

$$
\begin{array}{cccccc}
x & Tx & T^2x & \cdots & T^nx & \cdots \\
\downarrow & \downarrow & \downarrow & \downarrow & \downarrow & \downarrow \\
\psi(x) & S\left(\psi(x)\right) & S^2\left(\psi(x)\right) & \cdots & S^n\left(\psi(x)\right) & \cdots
\end{array}
$$

Definition 1.2.18. *Two dynamical systems* $(X, \mathcal{F}, \rho, T)$ *and* $(X', \mathcal{F}', \rho', T')$ *are isomorphic if there exist measurable sets* $N \subset X$ *and* $N' \subset$

X' with $\rho(X \setminus N) = \rho'(X' \setminus N') = 0$ and $T(N) \subset N$, $T'(N') \subset N'$, and finally if there exists a measurable map $\psi : N \to N'$ such that (i)–(iv) are satisfied.

Example 1.2.19. Let $X = [0, 1)$ and $Y := \{0, 1, \ldots, 9\}^{\mathbb{N}}$, the set of all sequences $(y_n)_{n \geq 1}$, with $y_n \in \{0, 1, \ldots, 9\}$ for $n \geq 1$. We now construct an isomorphism between $([0, 1), \mathcal{B}, \lambda, T)$ and $(Y, \mathcal{B}', \lambda', T')$, where $\mathcal{B}'$ is the σ-algebra generated by the cylinders, where λ' is defined on cylinders by

$$\lambda'(\{(y_i)_{i \geq 1} \in Y : y_1 = a_1, y_2 = a_2, \ldots, y_n = a_n\}) = \frac{1}{10^n},$$

and where T' is the *left shift*, given by

$$T'((a_1, a_2, a_3, \ldots)) := (a_2, a_3, \ldots),$$

for any $(a_1, a_2, a_3, \ldots) \in Y$. Notice that in everyday life we usually write $.a_1 a_2 a_3 \ldots$ instead of $(a_1, a_2, a_3, \ldots)$. ∎

Exercise 1.2.20. Show that the shift T' is λ'-measure preserving. ∎

Define $\psi : [0, 1) \to Y = \{0, 1, \ldots, 9\}^{\mathbb{N}}$ by

$$\psi : x = \sum_{k=1}^{\infty} \frac{a_k}{10^k} \mapsto (a_k)_{k \geq 1},$$

where $\sum_{k=1}^{\infty} a_k/10^k$ is the decimal expansion of x, and let

$$C(i_1, \ldots, i_n) = \{(y_i)_{i \geq 1} \in Y : y_1 = i_1, \ldots, y_n = i_n\}.$$

In order to see that ψ is an isomorphism one needs to verify measurability and measure preservingness on cylinders:

$$\psi^{-1}(C(i_1, \ldots, i_n))$$
$$= \left[\frac{i_1}{10} + \frac{i_2}{10^2} + \cdots + \frac{i_n}{10^n}, \frac{i_1}{10} + \frac{i_2}{10^2} + \cdots + \frac{i_n + 1}{10^n} \right)$$

and

$$\lambda \left(\psi^{-1}(C(i_1, \dots, i_n)) \right) = \frac{1}{10^n} = \lambda \left(\Delta(i_1, \dots, i_n) \right).$$

Note that

$$N = \{(y_i)_{i \geq 1} \in Y : \text{there exists a } k \geq 1 \text{ such that } y_i = 9 \text{ for all } i \geq k\}$$

is a subset of Y of measure 0. Setting $\tilde{Y} = Y \setminus N$, then $\psi : [0, 1) \to \tilde{Y}$ is a bijection, since every $x \in [0, 1)$ has a unique decimal expansion (generated by T; see Section 1.2.1). Finally, it is easy to see that $\psi \circ T = T' \circ \psi$. The dynamical system $(Y, \mathcal{B}', \lambda', T')$ is known as the Bernoulli shift on the symbols $0, 1, \dots, 9$, where each symbol comes with probability (*weight*) $\frac{1}{10}$.

In general a *Bernoulli shift* is defined as follows. Consider any probability space $(Y, \mathcal{F}, \mu)$. Let $(X, \mathcal{C}, \mu') = \prod_{n=1}^{\infty}(Y, \mathcal{F}, \mu)$, so that $X = Y^{\mathbb{N}}$ is the space of all one-sided sequences of elements of Y, and $\mathcal{C}$ is the σ-algebra generated by cylinders of the form

$$\{x = (x_k)_{k \geq 1} : x_i \in A_1, \dots, x_{i+n-1} \in A_n\}$$

where $A_1, \dots, A_n \in \mathcal{F}$ and $i, n \in \mathbb{N}$. Furthermore, μ' is the product measure defined on cylinders by

$$\mu'(\{x = (x_k)_{k \geq 1} : x_i \in A_1, \dots, x_{i+n-1} \in A_n\})$$
$$= \mu(A_1) \times \cdots \times \mu(A_n).$$

In case $Y = \{a_1, a_2, \dots\}$ is a finite or countable discrete space, we refer to μ' as the product measure with weights $\mu(C_1), \mu(C_2), \dots$, where $C_i = \{x = (x_k)_{k \geq 1} : x_1 = a_i\}$. Finally, define $T' : X \to X$ by $T'((y_n)_{n \geq 1}) = (x_n)_{n \geq 1}$, where $x_n = y_{n+1}$ for $n \in \mathbb{N}$. Of course, by the above notation we mean that X has the product structure. Any system isomorphic to the system $(X, \mathcal{C}, \mu', T')$ is measure preserving and is referred to as a *one-sided Bernoulli shift*. This definition can be naturally extended to a *two-sided Bernoulli shift* by replacing $\mathbb{N}$ by $\mathbb{Z}$.

Remark 1.2.21. In case T is defined on a complete probability space, which is isomorphic to the completion of a Bernoulli shift, then T is also called Bernoulli.   ∎

Exercise 1.2.22. Consider $\big([0, 1)^2, \mathcal{L} \times \mathcal{L}, \lambda \times \lambda\big)$, where $\mathcal{L} \times \mathcal{L}$ is the product Lebesgue σ-algebra, and $\lambda \times \lambda$ is the product Lebesgue measure. Let $T : [0, 1)^2 \to [0, 1)^2$ be given by

$$T(x, y) = \begin{cases} (2x, \tfrac{1}{2} y), & 0 \le x < \tfrac{1}{2} \\ (2x - 1, \tfrac{1}{2}(y + 1)), & \tfrac{1}{2} \le x < 1. \end{cases}$$

The map T is called the *Baker's transformation* (draw a picture to see that T moves $[0, 1)^2$ around similar to the way a baker folds a piece of dough).

(i) Use Lemma 1.2.14 to show that T is measurable and measure preserving with respect to $\lambda \times \lambda$, by checking this on rectangles of the form $(a, b) \times (c, d)$.

(ii) Show that T is isomorphic to the completion of the two-sided Bernoulli shift T' on $\big(\{0, 1\}^{\mathbb{Z}}, \mathcal{F}, \mu\big)$, where $\mathcal{F}$ is the σ-algebra generated by cylinders of the form

$$\Delta = \{x_{-k} = a_{-k}, \ldots, x_\ell = a_\ell : a_i \in \{0, 1\}, i = -k, \ldots, \ell\},$$

$$k, \ell \ge 0,$$

and μ the product measure with weights $(\tfrac{1}{2}, \tfrac{1}{2})$ (so $\mu(\Delta) = (\tfrac{1}{2})^{k+\ell+1}$).   ∎

1.3 Continued fractions

Let $x \in (0, 1)$; in this section we will see that one can write x as a continued fraction:

$$x = \cfrac{1}{a_1 + \cfrac{1}{a_2 + \cfrac{1}{a_3 + \cfrac{1}{\ddots}}}} \quad , \tag{1.6}$$

where $a_k \geq 1$. Obviously x is a rational number in case (1.6) is a finite expansion. Conversely, every rational number x has a finite continued fraction due to Euclid's algorithm, i.e., the division algorithm.

1.3.1 Euclid's algorithm

How is an expansion such as (1.6) generated? To answer this question we briefly review *Euclid's algorithm*, also known as the division algorithm. Let $a, b \in \mathbb{Z}$ and assume for convenience that $a > b > 0$. Put

$$r_0 := a, \quad r_1 := b,$$

and determine $a_1 \geq 1$, $r_2 \geq 0$, such that

$$r_0 = a_1 r_1 + r_2 ,$$

where $0 \leq r_2 < r_1$. In case $r_2 \neq 0$, we repeat this procedure, which clearly will stop after at most r_1 steps: There exists a positive integer n such that $r_n \neq 0$,

$$r_k = a_{k+1} r_{k+1} + r_{k+2} \quad \text{for} \quad k \leq n - 1$$

and

$$0 = r_{n+1} < r_n < \cdots < r_1.$$

Then, as is well known, we have that

$$(a, b) = r_n,$$

where (a, b) denotes the greatest common divisor (gcd) of a and b.

Exercise 1.3.1. Determine the gcd's $(13, 20)$, $(171951, 31250000)$ and $(5, 14)$.   ■

   Let us consider Euclid's algorithm more closely; obviously one has

$$a_1 = \left\lfloor \frac{r_0}{r_1} \right\rfloor, \; a_2 = \left\lfloor \frac{r_1}{r_2} \right\rfloor, \; \ldots, \; a_n = \left\lfloor \frac{r_{n-1}}{r_n} \right\rfloor,$$

where $\lfloor \xi \rfloor$ denotes the greatest integer not exceeding ξ. Putting

$$x = \frac{b}{a} = \frac{r_1}{r_0}, \; T_1 = \frac{r_2}{r_1}, \; T_2 = \frac{r_3}{r_2}, \ldots, \; T_{n-1} = \frac{r_n}{r_{n-1}},$$

one has

$$\frac{1}{x} = a_1 + T_1$$

$$\frac{1}{T_1} = a_2 + T_2$$

$$\vdots$$

$$\frac{1}{T_{n-2}} = a_{n-1} + T_{n-1}$$

$$\frac{1}{T_{n-1}} = a_n + 0,$$

and therefore one finds

$$x = \cfrac{1}{a_1 + T_1} = \cfrac{1}{a_1 + \cfrac{1}{a_2 + T_2}} = \cdots = \cfrac{1}{a_1 + \cfrac{1}{a_2 + \cdots + \cfrac{1}{a_n}}} \; . \quad (1.7)$$

An expression as in the right-hand side of (1.7) is called a *finite regular continued fraction*. It follows from Euclid's algorithm that each $x = \frac{p}{q} \in \mathbb{Q}$ can be written as a finite regular continued fraction

$$x = a_0 + \cfrac{1}{a_1 + \cfrac{1}{a_2 + \cfrac{\cdots}{} + \cfrac{1}{a_n}}} , \qquad (1.8)$$

where $a_0 \in \mathbb{Z}$ is such that $x - a_0 \in [0, 1)$. The right-hand side of (1.8) is abbreviated as

$$[\, a_0;\, a_1,\, a_2, \ldots ,\, a_n \,].$$

Euclid's algorithm yields that $a_n \geq 2$. Due to this, each rational number x has two regular continued fraction expansions, viz.

$$[\, a_0;\, a_1,\, a_2, \ldots ,\, a_{n-1},\, a_n \,] = [\, a_0;\, a_1,\, a_2, \ldots ,\, a_{n-1},\, a_n - 1,\, 1 \,].$$

In this case n is called the *depth* and $\epsilon(\frac{p}{q}) := (-1)^n$ is called the *signature* of $x = \frac{p}{q}$.

Exercise 1.3.2. Determine the continued fraction expansions of $\frac{13}{20}$, $\frac{5}{14}$ and $\frac{171951}{31250000}$. Compare the lengths of these continued fractions with the decimal expansions of these numbers as given in Section 1.1. ∎

Of course, there is no reason whatsoever to stick to rationals. We have the following definition.

Definition 1.3.3. *The regular continued fraction operator* T : $[0, 1) \to [0, 1)$ *is defined by*

$$Tx := \frac{1}{x} - \left\lfloor \frac{1}{x} \right\rfloor, \quad x \neq 0; \quad T0 := 0.$$

The map T is illustrated in Figure 1.3.

Now let $x \in \mathbb{R} \setminus \mathbb{Q}$ and, as in the rational case, let $a_0 \in \mathbb{Z}$ be such that $x - a_0 \in [0, 1)$. Putting

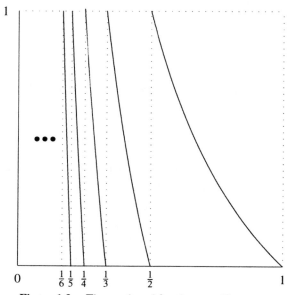

Figure 1.3.   The continued fraction map T

$$T_0 := x - a_0, \ T_1 := T(x - a_0), \ T_2 := T(T_1), \ldots ,$$

it follows at once from the above definition that

$$T_n \in [0, 1) \setminus \mathbb{Q}, \quad \text{for all } n \geq 0.$$

Moreover, setting

$$a_n = a_n(x) := \left\lfloor \frac{1}{T_{n-1}} \right\rfloor , \ n \geq 1,$$

one has

$$x = a_0 + \frac{1}{a_1 + T_1} = a_0 + \cfrac{1}{a_1 + \cfrac{1}{a_2 + T_2}} = \cdots$$

$$= a_0 + \cfrac{1}{a_1 + \cfrac{1}{a_2 + \cdots + \cfrac{1}{a_n + T_n}}} \tag{1.9}$$

$$= [\, a_0;\, a_1,\, a_2,\, \ldots,\, a_{n-1},\, a_n + T_n\,], \quad n \geq 1.$$

The integers $a_0, a_1, \ldots$ are called the digits or *partial quotients* of x and one has that $a_i \geq 1$ for $i \geq 1$. We denote the (regular) continued fraction expansion of x by

$$x = [\, a_0;\, a_1,\, a_2,\, \ldots,\, a_n,\, \ldots\,], \tag{1.10}$$

where this notation should be understood as

$$\lim_{n \to \infty} [\, a_0;\, a_1,\, a_2,\, \ldots,\, a_n + T_n\,].$$

That this limit exists, and that every irrational x has a unique (regular) continued fraction expansion, will be shown in the next subsection. In case x is a rational, applying T to the fractional part of x yields a finite continued fraction expansion, where the last partial quotient is greater than 1.

Truncating (1.10) yields the so-called (*regular*) *continued fraction convergents*

$$a_0 + \cfrac{1}{a_1 + \cfrac{1}{a_2 + \cdots + \cfrac{1}{a_n}}} = [\, a_0;\, a_1,\, \ldots,\, a_n\,]$$

of x. Clearly these convergents are rational numbers; see Exercise 1.3.8.

Example 1.3.4. Let $G = \frac{1}{2}(\sqrt{5} + 1)$ (the so-called *golden mean* or *golden ratio*). An easy calculation shows that $G^2 = G + 1$, from which it follows that

$$G = 1 + \frac{1}{G} = 1 + \cfrac{1}{1 + \cfrac{1}{G}}.$$

Thus we see that $G = [\,1;\ \overline{1}\,] = [\,1;\ 1, \dots,\ 1, \dots\,]$ (here—as usual—the bar indicates the period). Now let us calculate the continued fraction expansion of $\sqrt{2}$. Let $x = \sqrt{2} - 1$; then

$$x = \cfrac{1}{2 + x}$$

from which we see that $\sqrt{2} = [\,1;\ \overline{2}\,]$. ∎

Exercise 1.3.5. Argue why any real irrational number x with an eventually-periodic continued fraction expansion is the root of a polynomial of degree 2 (such an x is called a *quadratic irrational*). ∎

The converse of the statement also holds, but is much harder to prove. It is known as the Theorem of Lagrange (1770); see [RS92], pp. 40-41.

1.3.2 Basic properties and matrices

In this section we will derive a number of basic properties of continued fractions using 2×2 matrices. In fact, this matrix representation establishes the connection between continued fractions and (a part of) algebraic geometry, a connection that has been beautifully explained by C. Series in [Ser82] and [Ser85]. Let

$$A = \begin{bmatrix} a & b \\ c & d \end{bmatrix} \in \mathrm{SL}_2(\mathbb{Z})\,,$$

i.e., A has integer entries a, b, c and d, and $\det(A) \in \{-1, +1\}$. The letters SL in $\mathrm{SL}_2(\mathbb{Z})$ stand for *special linear*. Now define a map $A : \mathbb{R} \cup \{\infty\} \to \mathbb{R} \cup \{\infty\}$ by

$$A(x) := \frac{ax + b}{cx + d}\,,\ x \in \mathbb{R} \cup \{\infty\}\,.$$

Such a map is also known as a *Möbius transformation*. Notice that we use the same notation both for the matrix A and for its associated Möbius transformation.

Exercise 1.3.6. Show that if A, $B \in SL_2(\mathbb{Z})$, one has that

$$(AB)(x) = A(B(x)),$$

where AB is the usual matrix product of A and B. ∎

Let $x \in \mathbb{R}$ be an irrational number with continued fraction expansion $x = [a_0; a_1, \ldots, a_n, \ldots]$. Define for $n \geq 1$ matrices A_n and M_n by

$$A_0 := \begin{bmatrix} 1 & a_0 \\ 0 & 1 \end{bmatrix}, \quad A_n := \begin{bmatrix} 0 & 1 \\ 1 & a_n \end{bmatrix}, \quad (1.11)$$

$$M_n := A_0 A_1 \cdots A_n, \ n \geq 1.$$

Exercise 1.3.7. With A_n and M_n as before (considered as Möbius transformations), show that

$$M_n(0) = [a_0; a_1, \ldots, a_n]$$

and

$$A_1 A_2 \cdots A_n = (A_n A_{n-1} \cdots A_1)^T.$$ ∎

Exercise 1.3.8. Writing

$$M_n = \begin{bmatrix} r_n & p_n \\ s_n & q_n \end{bmatrix}, \ n \geq 0,$$

show that $(p_n, q_n) = 1$. Use $M_n = M_{n-1} A_n$ to show that

$$r_n = p_{n-1}, \quad s_n = q_{n-1},$$

$$p_{n-1} q_n - p_n q_{n-1} = (-1)^n, \quad n \geq 1, \quad \text{and}$$

$$\frac{p_n}{q_n} = [a_0; a_1, \ldots, a_n], \ n \geq 1.$$

Furthermore, show that the sequences $(p_n)_{n \geq -1}$ and $(q_n)_{n \geq -1}$ satisfy the following recurrence relations

$$p_{-1} := 1; \quad p_0 := a_0; \quad p_n = a_n p_{n-1} + p_{n-2}, \ n \geq 1,$$
$$q_{-1} := 0; \quad q_0 := 1; \quad q_n = a_n q_{n-1} + q_{n-2}, \ n \geq 1.$$
(1.12)

Finally, use (1.12) to show that $p_n(x) = q_{n-1}(Tx)$ for all $n \geq 0$, where $T_n = T^n x$; see Section 1.3.1. ∎

Exercise 1.3.9. Use the recurrence relation for the q_n's to show that

$$\frac{q_{n-1}}{q_n} = [0; a_n, \ldots, a_1].$$ ∎

As promised we will now show that $\lim_{n \to \infty} \frac{p_n}{q_n} = x$. To this end, we define one more matrix:

$$A_n^* := \begin{bmatrix} 0 & 1 \\ 1 & a_n + T_n \end{bmatrix}, \quad \text{for } n \geq 1.$$

Exercise 1.3.10. Show that $x = (M_{n-1} A_n^*)(0)$. Use the fact that

$$M_{n-1} = \begin{bmatrix} p_{n-2} & p_{n-1} \\ q_{n-2} & q_{n-1} \end{bmatrix}, \quad \text{for } n \geq 1,$$

and the above mentioned recurrence relations for $(p_n)_{n \geq -1}$ and $(q_n)_{n \geq -1}$ to show that

$$x = \frac{p_n + T_n p_{n-1}}{q_n + T_n q_{n-1}}, \quad \text{for } n \geq 1,$$

i.e., $x = M_n(T^n x)$. Finally, use the fact that $p_{n-1} q_n - p_n q_{n-1} = (-1)^n$ to conclude that

$$x - \frac{p_n}{q_n} = \frac{(-1)^n T_n}{q_n(q_n + T_n q_{n-1})}, \quad \text{for } n \geq 1. \tag{1.13}$$ ∎

Since $T_n \in [0, 1)$ we have that

$$\left| x - \frac{p_n}{q_n} \right| < \frac{1}{q_n^2}, \quad \text{for } n \geq 1. \tag{1.14}$$

The sequence $(q_n)_{n \geq 0}$ is a monotone increasing sequence of positive integers, which is—in case all a_i's are all equal to 1—the *Fibonacci sequence* $(\mathcal{F}_n)_{n \geq 1}$, given by

$$1, \ 1, \ 2, \ 3, \ 5, \ 8, \ 13, \ 21, \dots .$$

Now (1.14) yields that $\lim\limits_{n \to \infty} \dfrac{p_n}{q_n} = x$.

Exercise 1.3.11. Show that

$$\frac{p_0}{q_0} < \frac{p_2}{q_2} < \cdots < x < \cdots < \frac{p_3}{q_3} < \frac{p_1}{q_1}. \qquad \blacksquare$$

1.3.3 Lebesgue vs. Gauss measure

The map T does not preserve Lebesgue measure λ; e.g., one has (see also Figure 1.3) that

$$T^{-1}\left(0, \frac{1}{2}\right) = \bigcup_{n=1}^{\infty} \left(\frac{1}{n + \frac{1}{2}}, \frac{1}{n}\right),$$

and therefore

$$\lambda\left(T^{-1}\left(0, \frac{1}{2}\right)\right) = \sum_{n=1}^{\infty}\left(\frac{1}{n} - \frac{1}{n + \frac{1}{2}}\right) = \sum_{n=1}^{\infty}\left(\frac{\frac{1}{2}}{n(n + \frac{1}{2})}\right)$$

$$= \sum_{n=1}^{\infty}\left(\frac{1}{n(2n + 1)}\right)$$

$$= 2 - \log 4 = .613706\cdots \neq \frac{1}{2} = \lambda\left(0, \frac{1}{2}\right).$$

The following question is a natural one: *Does there exist a T-invariant measure μ equivalent to the Lebesgue measure λ?* I.e., does there exist a T-invariant measure μ that has the same sets of measure zero as λ?

Gauss [Gau76] found such an invariant measure in 1800, and this measure is known today as the *Gauss measure* μ, given by

$$\mu(A) = \frac{1}{\log 2} \int_A \frac{1}{1+x} dx$$

for all Lebesgue sets $A \subset [0, 1)$, where log refers to the natural logarithm. Nobody knows how Gauss found μ, and his achievement is even more remarkable if we realize that modern probability theory and ergodic theory started almost a century later! In general, finding the invariant measure is a difficult task.

Exercise 1.3.12. Use Figure 1.3 to show that for each interval $(a, b) \subset [0, 1)$ one has

$$T^{-1}(a, b) = \bigcup_{n=1}^{\infty} \left(\frac{1}{n+b}, \frac{1}{n+a} \right).$$

Then use this to prove that $\mu(a, b) = \mu(T^{-1}(a, b))$. As mentioned before, this implies that T is measure preserving. ∎

As in the case of the decimal map, we define cylinders $\Delta_n = \Delta(a_1, a_2, \ldots, a_n)$ by

$$\Delta(a_1, \ldots, a_n) := \{x \in [0, 1) : a_1(x) = a_1, \ldots, a_n(x) = a_n\},$$

where $a_j \in \mathbb{N}$ for each $1 \leq j \leq n$.

Exercise 1.3.13. Show that

$$\Delta(1) = \left(\frac{1}{2}, 1 \right) \quad \text{and that} \quad \Delta(1) = \left(\frac{1}{n+1}, \frac{1}{n} \right], \quad \text{for } n \geq 2.$$

Determine $\Delta(1, 1)$ and $\Delta(m, n)$, for $m, n \geq 1$. ∎

Exercise 1.3.14. Show that the above-defined cylinders Δ_n form a semi-algebra generating the Borel σ-algebra $\mathcal{B}$. ∎

Exercise 1.3.15. Show that $\Delta(a_1, a_2, \ldots, a_k)$ is an interval in $[0, 1)$ with endpoints

$$\frac{p_k}{q_k} \quad \text{and} \quad \frac{p_k + p_{k-1}}{q_k + q_{k-1}} \, ,$$

and conclude that

$$\lambda\left(\Delta(a_1, a_2, \ldots, a_k)\right) = \frac{1}{q_k(q_k + q_{k-1})} \, .$$

When is p_k/q_k the left-hand endpoint of $\Delta(a_1, a_2, \ldots, a_k)$?   ■

Exercise 1.3.16. Using Exercise 1.3.7, show that

$$\mu(\Delta(a_k, a_{k-1}, \ldots, a_1)) = \mu(\Delta(a_1, a_2, \ldots, a_k)) \, ,$$

where μ is the Gauss measure.   ■

1.4 For further reading

Although there are many introductory texts on both (elementary) number theory and measure theory/probability theory, only a few books have—in spirit—a considerable overlap with this chapter.

The first book that should be mentioned here is the famous classic, *An Introduction to the Theory of Numbers* by G.H. Hardy and E.M. Wright [HW79]. Although the first edition of this great book appeared in 1938, it is still very much worthwhile to study it. However, due to its old age, there is no mention of ergodic theory, which is one of our main subjects.

Another old book, but one very much in line with the spirit and subjects of our book, is P. Billingsley's *Ergodic Theory and Information* [Bil65]. The books by A.Ya. Khintchine [Khi63] and A.M. Rockett and P. Szüsz [RS92] both deal with continued fractions. Their results are largely different from those presented here, due to a different point of view. We want to show that there is a natural interaction between number theory and ergodic theory, while [Khi63] and [RS92] address the interaction between number theory (more specifically: the theory of continued fractions) and probability theory.

Other introductory books tend to focus on individual topics of this chapter, such as H. Davenport [Dav92] and H. Rademacher [Rad83] on elementary number theory, and W. Rudin [Rud87] and H.L. Royden [Roy88] on probability and measure.

Finally, F. Schweiger's *Ergodic Theory of Fibred Systems and Metric Number Theory* [Sch95] gives further back-ground information on all the subjects mentioned in this book—and many we left out!

CHAPTER **2**

Variations on a theme (Other expansions)

In this chapter we investigate how far the dynamics of the decimal expansion can be generalized without losing its 'essential properties', like Bernoullicity. We will see that the underlying property that guarantees, for instance, independence of the digits is that the map T generating the expansion is piecewise linear onto $[0, 1)$ on each partition element.

2.1 n-ary expansions

The ideas and results for the decimal expansion go through in exactly the same way if we look at n-ary expansions of numbers in $[0,1)$. Here, $n \geq 2$ is an integer and every irrational number $x \in [0, 1)$ can be uniquely written as

$$x = \sum_{k=1}^{\infty} \frac{a_k(x)}{n^k} \ , \quad a_k = a_k(x) \in \{0, 1, \ldots, n - 1\} .$$

Exercise 2.1.1. Let $n \in \mathbb{Z}$, $n \geq 2$, and let x be a real number with a finite or eventually-periodic n-ary expansion. Show that $x \in \mathbb{Q}$. ∎

One finds—among other things—the following proposition.

Proposition 2.1.2. *Let $n \in \mathbb{Z}$, $n \geq 2$, and let $x \in [0, 1)$. Then*

1. *x has a finite n-ary expansion if and only if there exist $p, q \in \mathbb{N}$, $(p, q) = 1$, $x = p/q$ and*

$$p_i | n \quad (\text{i.e., } p_i \text{ divides } n)$$

   for all primes p_i such that $p_i | q$.
2. *x has a purely-periodic n-ary expansion if and only if there exist $p, q \in \mathbb{N}$, $(p, q) = 1$, $x = p/q$ and $(q, n) = 1$.*

Decimal expansions are of course an example of n-ary expansions, which are generated by iterations of the map $Tx = nx \pmod 1$. That is, $Tx = nx - a_1$ where $a_1 = a_1(x)$ is such that $nx - a_1 \in [0, 1)$, or—equivalently—a_1 is such that $x \in [\frac{a_1}{n}, \frac{a_1+1}{n})$. From $Tx = nx - a_1$ we have that $x = \frac{a_1}{n} + \frac{Tx}{n}$. Putting $a_1(x) = \lfloor nx \rfloor$, $a_2(x) = \lfloor nTx \rfloor$, $\ldots$, $a_k(x) = \lfloor nT^{k-1}x \rfloor$, $\ldots$ we find

$$
\begin{aligned}
x &= \frac{a_1}{n} + \frac{Tx}{n} = \frac{a_1}{n} + \frac{a_2}{n^2} + \frac{T^2 x}{n^2} \\
&= \frac{a_1}{n} + \frac{a_2}{n^2} + \cdots + \frac{a_k}{n^k} + \frac{T^k x}{n^k} \\
&= \frac{a_1}{n} + \frac{a_2}{n^2} + \cdots + \frac{a_k}{n^k} + \cdots .
\end{aligned}
$$

Notice that if $(a, b) \subset [0, 1)$, then

$$T^{-1}(a, b) = \bigcup_{i=0}^{n-1} \left(\frac{i}{n} + \frac{a}{n}, \frac{i}{n} + \frac{b}{n} \right),$$

so $\lambda(T^{-1}(a, b)) = \sum_{i=0}^{n-1} (\frac{b}{n} - \frac{a}{n}) = b - a = \lambda(a, b)$, and T is measure preserving with respect to Lebesgue measure.

Exercise 2.1.3. Show that the digits of n-ary expansions are independent and identically distributed with the uniform distribution on the set

$\{0, 1, \ldots, n - 1\}$, i.e., show that

$$\lambda\{x \in [0, 1) : a_1(x) = i_1, \ldots, a_k(x) = i_k\} = \frac{1}{n^k},$$

for any $k \geq 1$ and $i_1, \ldots, i_k \in \{0, 1, \ldots, n - 1\}$. ∎

Nowadays bases 2 (binary expansion) and 10 (decimal expansion) are the most commonly used, but in former days bases like 6 (and its multiples 12 and 24) were immensely popular. In fact, traces of this can still be found in everyday life; only think of hours, minutes, seconds!

Exercise 2.1.4. Find other traces of bases different from 2 and 10 in everyday life. ∎

Exercise 2.1.5. Clearly base 2 is very handy in computer science, and base 10 is directly at hand. Think of reasons why base 6 instead of base 10 was chosen in former days (it was Napoleon Bonaparte who decreed the use of base 10 by law). ∎

Now suppose $x \in [0, 1)$ is an irrational number, and let

$$.a_1 a_2 \ldots a_k \ldots \tag{2.1}$$

be its expansion in base 2, while

$$.d_1 d_2 \ldots d_k \ldots \tag{2.2}$$

is the decimal expansion of x. In general, the infinite expansions (2.1) and (2.2) are not given to us, but only the 'finite truncations'

$$\alpha = .a_1 a_2 \ldots a_k \quad \text{and} \quad \beta = .d_1 d_2 \ldots d_k .$$

An important question is now: *Which of these two rational approximations α and β of x is closer to x? That is: Which of the two yields the better approximation?* Although this question cannot be answered without further information about x, one has the feeling β is—in gen-

eral (whatever that means)—closer to x than α is. One has the feeling
that β contains more information about x than α does. Perhaps this
is more obvious if we assume that (2.2) is the expansion of x in base
$10,000$. In that case one has

$$x \in \left[\beta, \beta + \frac{1}{10,000^k} \right) ,$$

and therefore

$$|x - \beta| < \frac{1}{10,000^k} .$$

Assuming that x can be found at some random location in the cylinder
set $[\beta, \beta + \frac{1}{10,000^k})$, one has that the expected distance of β to x, given
$d_1, \ldots, d_k$, equals

$$\frac{1}{2} \frac{1}{10,000^k} .$$

In the same way one finds that the expected distance of α to x, given
$a_1, \ldots, a_k$, equals

$$\frac{1}{2^{k+1}} .$$

So, in general there seems to be a relation between the amount of information contained in finite strings of an expansion and the size of the
base n. In order to get a grip on the amount of information, the notion
of *entropy* was introduced by Shannon [Sha48] in the 1940s in information theory and by A.N. Kolmogorov [Kol58] in 1958 in ergodic
theory; see also [Kol59]. Entropy is perhaps the single most important
notion for understanding the complexity of a system. We will return to
it in Chapter 6.

2.2 Lüroth series

In the n-ary case we had a partition of $[0, 1)$ with partition elements
of the same length and the transformation was linear with range $[0, 1)$

on each piece. We want to generalize these systems. First, let us drop
the condition of equal length. We now look at the case where we just
have a partition on $[0, 1)$ (countable or finite) and the transformation is
full on each partition element, i.e., it maps each partition element onto
$[0, 1)$.

Another kind of series expansion, introduced by J. Lüroth [Lür83]
in 1883, motivates this approach. Several authors have studied the
dynamics of such systems. Take as partition of $[0, 1)$ the intervals
$[\frac{1}{n+1}, \frac{1}{n})$ where $n \in \mathbb{N}$. Every number $x \in [0, 1)$ can be written as
a finite or infinite series, the so-called Lüroth (series) expansion

$$x = \frac{1}{a_1(x)} + \frac{1}{a_1(x)(a_1(x) - 1)a_2(x)} + \cdots$$

$$+ \frac{1}{a_1(x)(a_1(x) - 1) \cdots a_{n-1}(x)(a_{n-1}(x) - 1)a_n(x)} + \cdots ;$$

here $a_k(x) \geq 2$ for each $k \geq 1$. How is such a series generated?

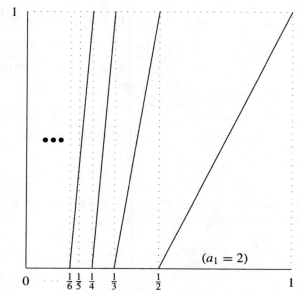

Figure 2.1.   The Lüroth Series map T

Let $T : [0, 1) \to [0, 1)$ be defined by

$$Tx = \begin{cases} n(n+1)x - n, & x \in [\frac{1}{n+1}, \frac{1}{n}), \\ 0, & x = 0. \end{cases} \qquad (2.3)$$

Let $x \neq 0$, for $k \geq 1$ and $T^{k-1}x \neq 0$ we define the digits $a_n = a_n(x)$ by

$$a_k(x) = a_1(T^{k-1}x),$$

where $a_1(x) = n$ if $x \in [\frac{1}{n}, \frac{1}{n-1})$, $n \geq 2$. Now (2.3) can be written as

$$Tx = \begin{cases} a_1(x)(a_1(x)-1)x - (a_1(x)-1), & x \neq 0, \\ 0, & x = 0. \end{cases}$$

Thus[1], for any $x \in (0, 1)$ such that $T^{k-1}x \neq 0$, we have

$$x = \frac{1}{a_1} + \frac{Tx}{a_1(a_1-1)} = \frac{1}{a_1} + \frac{1}{a_1(a_1-1)}\left(\frac{1}{a_2} + \frac{T^2x}{a_2(a_2-1)}\right)$$

$$= \frac{1}{a_1} + \frac{1}{a_1(a_1-1)a_2} + \frac{T^2x}{a_1(a_1-1)a_2(a_2-1)}$$

$$\vdots$$

$$= \frac{1}{a_1} + \cdots + \frac{1}{a_1(a_1-1)\cdots a_{k-1}(a_{k-1}-1)a_k}$$

$$+ \frac{T^kx}{a_1(a_1-1)\cdots a_k(a_k-1)}.$$

Notice that, if $T^{k-1}x = 0$ for some $k \geq 1$, and if we assume that k is the smallest positive integer with this property, then

$$x = \frac{1}{a_1} + \cdots + \frac{1}{a_1(a_1-1)\cdots a_{k-1}(a_{k-1}-1)a_k}.$$

---

[1]For ease of notation we drop the argument x from the functions $a_k(x)$.

In case $T^{k-1}x \neq 0$ for all $k \geq 1$, one gets

$$x = \frac{1}{a_1} + \frac{1}{a_1(a_1 - 1)a_2} + \cdots + \frac{1}{a_1(a_1 - 1) \cdots a_{k-1}(a_{k-1} - 1)a_k} + \cdots,$$

where $a_k \geq 2$ for each $k \geq 1$. Let us convince ourselves that this last infinite series indeed converges to x. Let $S_k = S_k(x)$ be the sum of the first k terms of the sum. Then

$$|x - S_k| = \left| \frac{T^k x}{a_1(a_1 - 1) \cdots a_k(a_k - 1)} \right|;$$

since $T^k x \in [0, 1)$ and $a_k \geq 2$ for all x and all $k \geq 1$, we find

$$|x - S_k| \leq \frac{1}{2^k} \to 0 \text{ as } k \to \infty.$$

From the above we also see that if x and y have the same Lüroth expansion, then, for each $k \geq 1$,

$$|x - y| \leq \frac{1}{2^{k-1}}$$

and it follows that x equals y.

As usual we consider Lebesgue measure λ on $[0, 1)$. In the next exercise you will show that T is measure preserving with respect to λ.

Exercise 2.2.1. Let $(a, b) \subset [0, 1)$. Show that

$$T^{-1}(a, b) = \bigcup_{k=2}^{\infty} \left(\frac{1}{k} + \frac{a}{k(k - 1)}, \frac{1}{k} + \frac{b}{k(k - 1)} \right),$$

and conclude from this that $\lambda(T^{-1}(a, b)) = \lambda(a, b)$.   ∎

Let us study the distribution of the digits

$$a_1, a_2 = a_1 \circ T, \ldots, a_k = a_1 \circ T^{k-1}, \ldots.$$

As before we define

$$\Delta(i) = \{x : a_1(x) = i\} \qquad \text{and}$$

$$\Delta(i_1, i_2, \ldots, i_k) = \{x : a_1(x) = i_1, a_2(x) = i_2, \ldots, a_k(x) = i_k\}.$$

Now if ≥ 2, then

$$\lambda\big(\Delta(a)\big) = \lambda(\{x : a_1(x) = a\}) = \lambda(T^{-(n-1)}(\Delta(a)))$$

$$= \lambda(\{x : a_1(T^{n-1}x) = a_n(x) = a\}) \qquad (2.4)$$

$$= \lambda((\frac{1}{a}, \frac{1}{a-1}]) = \frac{1}{a(a-1)} \,.$$

Therefore, $a_1(x)$, $a_2(x)$, ... are identically distributed random variables with distribution as given in (2.4).
Now consider

$$\lambda\,(\Delta(a_1, a_2, \dots, a_k))$$
$$= \lambda\,(\{x : a_1(x) = a_1, a_2(x) = a_2, \dots, a_k(x) = a_k\})$$
$$= \lambda\,\big(\{x : a_{j+1}(x) = a_1, \dots, a_{j+k}(x) = a_k\}\big)$$
$$= \lambda\,(\text{all } x \text{ whose Lüroth series begins with } P_k/Q_k)\,,$$

where

$$P_k/Q_k = \frac{1}{a_1} + \frac{1}{a_1(a_1-1)a_2} + \cdots + \frac{1}{a_1(a_1-1)\cdots a_{k-1}(a_{k-1}-1)a_k}\,.$$

Exercise 2.2.2. Show that $\Delta(a_1, a_2, \dots, a_k)$ is an interval in $[0, 1)$ with endpoints

$$\frac{P_k}{Q_k} \quad \text{and} \quad \frac{P_k}{Q_k} + \frac{1}{a_1(a_1-1)\cdots a_k(a_k-1)}\,,$$

and conclude that

$$\lambda\,(\Delta(a_1, a_2, \dots, a_k)) = \prod_{i=1}^{k} \lambda(\{x \in [0, 1) : a_i(x) = a_i\})\,.$$

Using that for any $k \geq 1$,

$$\sum_{a_1=2}^{\infty} \cdots \sum_{a_k=2}^{\infty} \frac{1}{a_1(a_1-1)\cdots a_k(a_k-1)} = 1\,,$$

show that for any $i_1 < i_2 < \cdots < i_j$,

$$\lambda \left(\{ x \in [0, 1) : a_{i_1}(x) = a_1, \, a_{i_2}(x) = a_2, \ldots, \, a_{i_j}(x) = a_j \} \right)$$

$$= \prod_{m=1}^{j} \lambda(\{ x \in [0, 1) : a_{i_m}(x) = a_m \}) \, . \qquad \blacksquare$$

Thus the digits functions $a_1(x)$, $a_2(x)$, ... are independent and identically distributed.

Exercise 2.2.3. Among the things that Lüroth showed in [Lür83] is that each rational has either a finite or a periodic expansion. Try to find a proof of this yourself by first showing that $T(\frac{p}{q}) = \frac{p'}{q}$ with $0 \le p' < q$. Conclude that there exist $0 \le m < n < q$ such that $T^m(\frac{p}{q}) = T^n(\frac{p}{q})$.

$\qquad \blacksquare$

Exercise 2.2.4. Show that we can identify the dynamical system $((0, 1], \mathcal{B}, \lambda, T)$ with a Bernoulli shift $(\{2, 3, \ldots\}^{\mathbb{N}}, \mathcal{F}, \mu, S)$, where $\mathcal{F}$ is the σ-algebra generated by the cylinders, S is the left-shift, and μ is the product measure with weights

$$\frac{1}{1 \times 2}, \, \frac{1}{2 \times 3}, \, \frac{1}{3 \times 4}, \ldots,$$

via

$$x = \frac{1}{a_1} + \frac{1}{a_1(a_1 - 1)a_2} + \cdots \; \mapsto \; [a_1, a_2, \ldots] \, . \qquad \blacksquare$$

2.3   Generalized Lüroth series

We will use the same dynamical mechanism that generated the n-ary expansions and the Lüroth series to define a family of series expansions, the so-called *generalized Lüroth series*, in short: GLS. We will see that these generalized Lüroth series have the same dynamical properties as the aforementioned n-ary expansions and the Lüroth series;

viz. the digits are independent and identically distributed, and the invariant measure of the underlying transformation is again Lebesgue
measure.

2.3.1 Introduction

Consider any partition $\mathcal{I} = \{[\ell_n, r_n) : n \in \mathcal{D}\}$ of $[0, 1)$ where $\mathcal{D} \subset \mathbb{Z}^+$
is finite or countable and $\sum_{n \in \mathcal{D}}(r_n - \ell_n) = 1$. We write $L_n = r_n - \ell_n$
and $I_n = [\ell_n, r_n)$ for $n \in \mathcal{D}$. Moreover, we assume that $i, j \in \mathcal{D}$ with
$i > j$ satisfy $0 < L_i \le L_j < 1$. $\mathcal{D}$ is called the digit set; see also
Figure 2.2.

$$\begin{array}{ccccccc} 0 & \ell_3\ r_3 & & \ell_1 & r_1 & \ell_2 & r_2 & 1 \end{array}$$

Figure 2.2. The partition $\mathcal{I}$

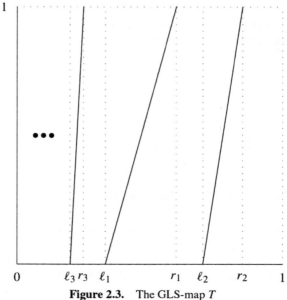

Figure 2.3. The GLS-map T

We will consider the following transformation T on $[0, 1)$:

$$Tx = \begin{cases} \dfrac{1}{r_n - \ell_n}x - \dfrac{\ell_n}{r_n - \ell_n}, & x \in I_n, \ n \in \mathcal{D}, \\[2mm] 0, & x \in I_\infty = [0, 1) \setminus \bigcup_{n \in \mathcal{D}} I_n ; \end{cases} \tag{2.5}$$

see also Figure 2.3.

Proposition 2.3.1. *The transformation T is measure preserving with respect to λ.*

Proof. To show that T is measure preserving, by Remark 1.2.15 it is enough to verify this on the open intervals. Let $0 \le a < b \le 1$; then

$$T^{-1}(a, b)$$

$$= \left(T^{-1}(a, b) \cap \bigcup_n I_n \right) \cup \left(T^{-1}(a, b) \cap I_\infty \right)$$

$$= \bigcup_n ((r_n - \ell_n)a + \ell_n, (r_n - \ell_n)b + \ell_n) \cup \left(T^{-1}(a, b) \cap I_\infty \right).$$

Since $\lambda(I_\infty) = 0$, it follows that

$$\lambda\left(T^{-1}(a, b) \right) = \sum_n (r_n - \ell_n)(b - a) = b - a = \lambda(a, b). \qquad \blacksquare$$

We want to iterate T in order to generate a series expansion of points x in $[0, 1)$, in fact of points x whose T-orbit never hits I_∞. We will show that the set of such points has measure 1.

We first need some notation. For $x \in [\ell_n, r_n)$, $n \in \mathcal{D}$, we write

$$s(x) = \frac{1}{r_n - \ell_n} \quad \text{and} \quad h(x) = \frac{\ell_n}{r_n - \ell_n},$$

so that $Tx = xs(x) - h(x)$. Now let

$$s_k(x) = \begin{cases} s(T^{k-1}x), & \text{if } T^{k-1}x \in \bigcup_{n \in \mathcal{D}} I_n, \\ \infty, & \text{otherwise,} \end{cases}$$

$$h_k(x) = \begin{cases} h(T^{k-1}x), & \text{if } T^{k-1}x \in \bigcup_{n \in \mathcal{D}} I_n, \\ 1, & \text{otherwise} \end{cases}$$

(thus $s(x) = s_1(x)$, $h(x) = h_1(x)$). From these definitions we see that for $x \in \bigcup_{n \in \mathcal{D}} I_n \cap (0, 1)$ such that $T^k x \in \bigcup_{n \in \mathcal{D}} I_n \cap (0, 1)$ for all $k \geq 1$, one has

$$x = \frac{h_1(x)}{s_1(x)} + \frac{Tx}{s_1(x)} = \frac{h_1}{s_1} + \frac{Tx}{s_1}$$

$$= \frac{h_1}{s_1} + \frac{1}{s_1}\left(\frac{h_2}{s_2} + \frac{T^2x}{s_2}\right) = \frac{h_1}{s_1} + \frac{h_2}{s_1 s_2} + \frac{T^2 x}{s_1 s_2}$$

$$= \frac{h_1}{s_1} + \frac{h_2}{s_1 s_2} + \cdots + \frac{h_k}{s_1 s_2 \cdots s_k} + \frac{T^k x}{s_1 s_2 \cdots s_k}$$

$$= \frac{h_1}{s_1} + \frac{h_2}{s_1 s_2} + \cdots + \frac{h_k}{s_1 s_2 \cdots s_k} + \cdots .$$

We refer to the above expansion as the GLS($\mathcal{I}$) expansion of x with a specified digit set $\mathcal{D}$. Such an expansion converges to x. Moreover, it is unique.

To prove the first statement we define the nth GLS-convergent P_k/Q_k of x by

$$\frac{P_k}{Q_k} = \frac{h_1}{s_1} + \frac{h_2}{s_1 s_2} + \cdots + \frac{h_k}{s_1 s_2 \cdots s_k};$$

then

$$\left| x - \frac{P_k}{Q_k} \right| = x - \frac{P_k}{Q_k} = \frac{T^k x}{s_1 s_2 \cdots s_k}. \tag{2.6}$$

Notice that

$$\frac{1}{s_k} = \text{length of the interval that } T^{k-1}x \text{ belongs to.}$$

For the proof of the second statement, use (2.6) and the triangle inequality.

Exercise 2.3.2. Let $L := \max_{n \in \mathcal{D}} L_n$. Show that

$$\left| x - \frac{P_k}{Q_k} \right| \le L^k \to 0 \quad \text{as} \quad k \to \infty .$$ ∎

Proposition 2.3.3. *Let J consist of all points in $(0, 1)$ with infinite GLS expansion; then J is T-invariant, $\lambda(J) = 1$ and $\lambda(I_\infty) = 0$.*

Proof. We only need to show that $\lambda(J) = 1$. Observe that $x \in (0, 1)$ has an infinite expansion, if and only if

$$T^k x \in \bigcup_{n \in \mathcal{D}} I_n \cap (0, 1) \quad \text{for all} \ k \ge 0.$$

This shows that

$$J = \bigcap_{k=0}^{\infty} T^{-k} \left(\bigcup_{n \in \mathcal{D}} I_n \cap (0, 1) \right).$$

Since T is measure preserving, and $\lambda \left(\bigcup_{n \in \mathcal{D}} I_n \cap (0, 1) \right) = 1$, we have

$$\lambda \left(T^{-k} \left(\bigcup_{n \in \mathcal{D}} I_n \cap (0, 1) \right) \right) = 1 \quad \text{for all} \ k \ge 0.$$

Hence, $\lambda(J) = 1$.   ∎

Notice that $x \in (0, 1)$ has a finite GLS($\mathcal{I}$) expansion of the form

$$x = \frac{h_1}{s_1} + \frac{h_2}{s_1 s_2} + \cdots + \frac{h_k}{s_1 s_2 \cdots s_k},$$

in case k is the least positive integer such that $T^k x = 0$, and $T^{k-1} x \notin I_\infty$.

Examples 2.3.4.

(1) *Decimal expansion:*

$$\mathcal{D} = \{0, 1, \ldots, 9\}, \quad I_k = \left[\frac{k}{10}, \frac{k+1}{10} \right), k \in \mathcal{D}$$

and $s_1(x) = 10$ which implies that $s_n(x) = 10$ for all x. Finally $h_1(x) = k$ if $x \in I_k$.

(2) *Lüroth series:*

$$\mathcal{D} = \{2, 3, \ldots\}, \quad I_k = \left[\frac{1}{k}, \frac{1}{k-1} \right), k \in \mathcal{D},$$

$s_1(x) = k(k-1) = a_1(a_1 - 1)$ if $x \in I_k$ and $s_k = a_k(a_k - 1)$, $h_k = a_k - 1$. Notice that

$$\frac{h_k}{s_1 s_2 \cdots s_k} = \frac{1}{a_1(a_1 - 1) \cdots a_{k-1}(a_{k-1} - 1)a_k} . \qquad \blacksquare$$

2.3.2   Digits

A GLS expansion is identified with the partition $\mathcal{I}$ and the index (or digit) set $\mathcal{D}$. Let x have an infinite GLS($\mathcal{I}$) expansion, given by

$$x = \frac{h_1}{s_1} + \frac{h_2}{s_1 s_2} + \cdots + \frac{h_k}{s_1 s_2 \cdots s_k} + \cdots .$$

Now h_k and s_k are identified once we know in which partition element $T^{k-1}x$ lies (h_k and s_k are constants determined by partition elements). Therefore, to determine the GLS-expansion of x (for a given $\mathcal{I}$ and $\mathcal{D}$) we only need to keep track of which partition elements the orbit of x visits. For $x \in [0, 1)$ we define the sequence of digits $a_n = a_n(x)$, $n \geq 1$, as follows

$$a_n = k \iff T^{n-1}x \in I_k , \quad k \in \mathcal{D} \cup \{\infty\}.$$

Thus the values of the digits of points $x \in [0, 1)$ are elements of $\mathcal{D}$; this is why $\mathcal{D}$ was called the digit set.

Notice that every GLS expansion determines a unique sequence of digits, and conversely. So

$$x = \sum_{k=1}^{\infty} \frac{h_k}{s_1 s_2 \cdots s_k} =: [a_1, a_2, \ldots] .$$

We can now define fundamental intervals (or cylinder sets) in the usual way. Setting

$$\Delta(i) = \{x : a_1(x) = i\} \text{ if } i \in \mathcal{D} \cup \{\infty\},$$

then

$$\Delta(i) = [l_i, r_i) \text{ if } i \in \mathcal{D}, \text{ and } \Delta(\infty) = I_\infty.$$

For $i_1, i_2, \dots, i_n \in \mathcal{D} \cup \{\infty\}$, define

$$\Delta(i_1, i_2, \dots, i_n) = \{x : a_1(x) = i_1, a_2(x) = i_2, \dots, a_n(x) = i_n\}.$$

Notice that, if $i_j = \infty$ for some $1 \leq j \leq n$, then $\Delta(i_1, i_2, \dots, i_n)$ is a subset of a set of measure zero, namely the set consisting of all points in $(0, 1)$ whose orbit hits I_∞.

Let us determine the cylinder sets $\Delta(i_1, \dots, i_k)$, for $i_1, i_2, \dots, i_n \in \mathcal{D}$. All points x with the same first k digits have the same first k terms in their GLS expansion. Let us call the sum of the first k terms p_k/q_k; then

$$x = \frac{p_k}{q_k} + \frac{T^k x}{s_1 \cdots s_k},$$

where $s_j = 1/L_{i_j}$ and $T^k x$ can vary freely in $[0, 1)$. This implies that

$$\Delta(i_1, \dots, i_k) = \left[\frac{p_k}{q_k}, \frac{p_k}{q_k} + \frac{1}{s_1 \cdots s_k}\right),$$

from which we clearly have

$$\lambda(\Delta(i_1, \dots, i_k)) = \frac{1}{s_1 \cdots s_k}.$$

Since $L_{i_j} = 1/s_j$ for each j, we find that

$$\lambda(\Delta(i_1, \dots, i_k)) = L_{i_1} L_{i_2} \cdots L_{i_k} = \lambda(\Delta(i_1))\lambda(\Delta(i_2)) \cdots \lambda(\Delta(i_k)).$$

Hence the digits are independent.

Exercise 2.3.5. Show that the above defined GLS-transformation T on $[0, 1)$ is isomorphic to the Bernoulli shift $(\mathcal{D}^{\mathbb{N}}, \mathcal{F}, \lambda', T')$ where T' is the left shift, $\mathcal{F}$ is the σ-algebra generated by the cylinder sets and λ' is the product measure giving the symbol $i \in \mathcal{D}$ weight L_i. ■

2.3.3   The general GLS case

So far, all the lines for our GLS-transformations had positive slopes. If we extend our domain to $[0, 1]$, we see that negative slopes work in the same way; what is important is that one has a countable partition of $[0, 1)$ as before, and that each partition element is mapped bijectively onto $(0, 1]$ or $[0, 1)$ by a linear map, with either positive or negative slope. We will briefly outline the construction; for more details, see [BBDK96].

We use the same notation as in Section 2.3.1, so the map T on $[0, 1)$ is as given in (2.5). We define the map $S \; : \; [0, 1] \to [0, 1]$ by

$$Sx := \begin{cases} \dfrac{r_n - x}{r_n - \ell_n}, & x \in I_n, \; n \in \mathcal{D}, \\ 0, & x \in (0, 1] \setminus \bigcup_{n \in \mathcal{D}} I_n, \\ 1, & x = 0. \end{cases}$$

Now let $\varepsilon = (\varepsilon(n))_{n \in \mathcal{D}}$ be an arbitrary, fixed sequence of zeros and ones (this sequence tells you whether you are going to use a map with positive or negative slope on the nth partition element). For $x \in (0, 1)$, let

$$\varepsilon(x) := \begin{cases} \varepsilon(n), & x \in I_n, \\ 0, & x \in (0, 1) \setminus \bigcup_{n \in \mathcal{D}} I_n, \end{cases}$$

and set $\varepsilon(0) = 0$ and $\varepsilon(1) = 1$. Define the map $T_\varepsilon \; : \; [0, 1] \to [0, 1]$ by

$$T_\varepsilon x := \varepsilon(x)Sx + (1 - \varepsilon(x))Tx, \quad x \in [0, 1]. \tag{2.7}$$

Let

$$s(x) := \frac{1}{r_n - \ell_n} \quad \text{and} \quad h(x) := \frac{\ell_n}{r_n - \ell_n}, \quad \text{in case } x \in I_n.$$

For $k \geq 1$, set $\varepsilon_k := \varepsilon(T_\varepsilon^{k-1}x)$,

$$s_k = s_k(x) := \begin{cases} s(T_\varepsilon^{k-1}x) & T_\varepsilon^{k-1}x \in \bigcup_n I_n \cap (0,1), \\ \infty & T_\varepsilon^{k-1}x \in [0,1] \setminus \bigcup_n I_n, \end{cases}$$

and h_k is defined in a similar way. Now, for

$$x \in \Omega = \bigcap_{k=0}^{\infty} \bigcup_{n \in \mathcal{D}} T_\varepsilon^{-k} (I_n \cap (0,1)),$$

one has

$$x = \frac{h_1 + \varepsilon_1}{s_1} + \frac{(-1)^{\varepsilon_1}}{s_1} T_\varepsilon x = \frac{h_1 + \varepsilon_1}{s_1} + (-1)^{\varepsilon_1} \frac{h_2 + \varepsilon_2}{s_1 s_2} + \cdots$$

$$+ (-1)^{\varepsilon_1 + \cdots + \varepsilon_{k-1}} \frac{h_k + \varepsilon_k}{s_1 s_2 \cdots s_k} + \frac{(-1)^{\varepsilon_1 + \cdots + \varepsilon_k}}{s_1 s_2 \cdots s_k} T_\varepsilon^k x .$$

For each $k \geq 1$ and $1 \leq i \leq k$ one has $s_i \geq 1/L > 1$, where $L = \max_{n \in \mathcal{D}} L_n$, and $T_\varepsilon^k x \leq 1$.

Exercise 2.3.6. Show that

$$\left| x - \frac{p_k}{q_k} \right| = \frac{T_\varepsilon^k x}{s_1 s_2 \cdots s_k} \leq L^k \to 0 \quad \text{as} \quad k \to \infty,$$

where

$$\frac{p_k}{q_k} = \frac{h_1 + \varepsilon_1}{s_1} + (-1)^{\varepsilon_1} \frac{h_2 + \varepsilon_2}{s_1 s_2} + \cdots + (-1)^{\varepsilon_1 + \cdots + \varepsilon_{k-1}} \frac{h_k + \varepsilon_k}{s_1 s_2 \cdots s_k} .$$

∎

Let $\varepsilon_0 := 0$; then for each $x \in \Omega$ one has

$$x = \sum_{n=1}^{\infty} (-1)^{\varepsilon_0 + \cdots + \varepsilon_{n-1}} \frac{h_n + \varepsilon_n}{s_1 s_2 \cdots s_n} . \tag{2.8}$$

Exercise 2.3.7. Show that T_ε is measure preserving with respect to λ, and $\lambda(\Omega) = 1$.

∎

For each $x \in [0, 1]$ we define its sequence of digits $a_k = a_k(x)$, $k \geq 1$, as follows:

$$a_k = n \iff T_\varepsilon^{k-1} x \in I_n \,,$$

for $n \in \mathcal{D} \cup \{\infty\}$, where $I_\infty = [0, 1] \setminus \bigcup_n I_n$. The expansion (2.8) is called the $(\mathcal{I}, \varepsilon)-$*Generalized Lüroth Series* (GLS) of x. Notice that for each $x \in \Omega$ there is a unique expansion (2.8), and therefore a unique sequence of digits $a_k \in \mathcal{D}$. Conversely, each sequence of digits a_k, $k \geq 1$, with $a_k \in \mathcal{D}$ defines a unique series expansion (2.8). We also write

$$x = \begin{bmatrix} \varepsilon_1, & \varepsilon_2, & \varepsilon_3, & \dots & \varepsilon_k, & \dots \\ a_1, & a_2, & a_3, & \dots & a_k, & \dots \end{bmatrix} .$$

Examples 2.3.8.

(1) *Alternating Lüroth*: Let $I_n := [\frac{1}{n}, \frac{1}{n-1})$, $n \geq 2$. In case $\varepsilon(n) = 0$ for $n \geq 2$, one gets the classical Lüroth Series, while $\varepsilon_n = 1$ for $n \geq 2$ yields the alternating Lüroth series; see also [KKK90] and [KKK91].

(2) *Tent map*: For $n \in \mathbb{N}, n \geq 2$, put $I_i = [\frac{i}{n}, \frac{i+1}{n})$, $i = 0, 1, \dots, n-1$. In case $\varepsilon(i) = 0$ for all i, the restriction of T_ε on $[0, 1)$ yields the n-ary expansion. In case $n = 2$ and $\varepsilon(0) = 0 = 1 - \varepsilon(1)$, T_ε is the tent map. ∎

Exercise 2.3.9. Let T_ε be an $(\mathcal{I}, \varepsilon)$-GLS map on $[0, 1]$, with digit set $\mathcal{D}$ and partition set $\mathcal{I}$. Let $a \in \mathcal{D}$; then we define the *jump transformation* S on $[0, 1]$ as follows

$$S(x) := \begin{cases} T_\varepsilon^k(x), & k(x) \in \mathbb{N} \\ 0, & k(x) = \infty \,, \end{cases}$$

where $k(x) := \inf\{m \in \mathbb{N}; \ a_m(x) = a\}$. Show that S is also a GLS map on $[0, 1]$, and find its GLS partition. (In case $\varepsilon(n) = 0$ for all $n \in \mathcal{D}$, replace $[0, 1]$ by $[0, 1)$). ∎

2.4 β-expansions

In this section we give examples of expansions T generated in a way similar to the GLS with positive slope, but where the 'shape' of T is changed. We will see that Lebesgue measure is no longer the invariant measure. These examples will be studied in detail in Chapters 3 and 4. See also Chapter 4 for more examples.

2.4.1 Lebesgue is no longer invariant

Let $\beta > 1$ be a real number. Consider the transformation $T_\beta : [0, 1) \to [0, 1)$, given by

$$T_\beta x := \beta x \bmod 1 .$$

If $\beta = n \in \mathbb{Z}$, we get the usual n-ary expansions, which have already been considered. Here we will concentrate on $\beta \notin \mathbb{Z}$.

For $x \in [0, 1)$ we write

$$d_1 = d_1(x) := \lfloor \beta x \rfloor \quad \text{and} \quad d_n = d_n(x) = d_1(T_\beta^{n-1}x) := \lfloor \beta T_\beta^{n-1}x \rfloor ,$$

and from $T_\beta x = \beta x - d_1$, $T_\beta^2 x = \beta T_\beta x - d_2, \ldots$ we see

$$x = \frac{d_1}{\beta} + \frac{T_\beta x}{\beta} = \frac{d_1}{\beta} + \frac{d_2}{\beta^2} + \cdots .$$

We call $d_n = \lfloor \beta T_\beta^{n-1}x \rfloor$, $n \geq 1$, the *digits* of the β-expansion of x. One clearly has $d_n \in \{0, 1, \ldots, \lfloor \beta \rfloor\}$.

Exercise 2.4.1. Let $\beta = G$, where $G = \frac{1}{2}(\sqrt{5} + 1) = 1.618\ldots$ is the 'golden mean' (see also Example 1.3.4 and Figure 3.1, p. 78), and consider the β-transformation $T_G(x) = Gx \bmod 1$.

(a) Let $x \in [0, 1)$. Show that $d_1(x)$ takes only two values: 0 or 1, and that $d_1(x) = 0 \iff x \in [0, g)$, where $g = G^{-1} = \frac{1}{2}(\sqrt{5} - 1) = G - 1$.

(b) Show that for every $x \in [0, 1)$ a digit 0 can be followed by a 0 or a 1, but that a digit 1 must be followed by a 0. ∎

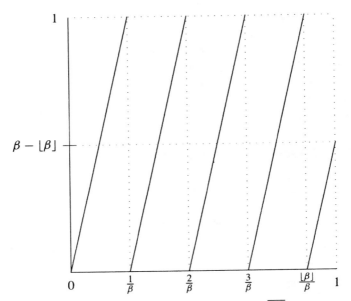

Figure 2.4.   The β-map T_β with $\beta = 4.\overline{504}$

If—as usual—λ denotes Lebesgue measure on $[0, 1]$, then T_β does not preserve λ if $\beta \notin \mathbb{N}$; for if a and b are in $[0, 1)$ such that $\beta - \lfloor \beta \rfloor < a < b < 1$, then

$$T_\beta^{-1}(a, b) = \bigcup_{j=0}^{\lfloor \beta \rfloor - 1} \left(\frac{a}{\beta} + \frac{j}{\beta}, \frac{b}{\beta} + \frac{j}{\beta} \right)$$

(see Figure 2.4), and therefore

$$\lambda(T_\beta^{-1}(a, b)) = \sum_{j=0}^{\lfloor \beta \rfloor - 1} \frac{b - a}{\beta} = \frac{\lfloor \beta \rfloor}{\beta}(b - a) .$$

Since $\beta \notin \mathbb{N}$ we have $\lfloor \beta \rfloor / \beta < 1$, and thus we find $\lambda(T_\beta^{-1}(a, b)) < \lambda(a, b)$.

A natural question is now: *Does there exist a measure ν of the form $\nu_\beta(A) = \int_A g_\beta(x)dx$, where g_β satisfies $0 < g_\beta(x) < \infty$,* *and* $v_\beta(T_\beta^{-1}A) = v_\beta(A)$? (See Section 3.1.1 for a definition of $\int_A g_\beta(x)dx$.) Such a measure v_β is said to be *equivalent* to Lebesgue measure λ; they share the same sets of measure zero.

Exercise 2.4.2. As in Exercise 2.4.1, let $\beta = G$, and consider the piecewise constant function g_G, given by

$$g_G(x) = \begin{cases} \dfrac{5 + 3\sqrt{5}}{10}, & 0 \le x < \dfrac{\sqrt{5}-1}{2}, \\[4mm] \dfrac{5 + \sqrt{5}}{10}, & \dfrac{\sqrt{5}-1}{2} \le x < 1. \end{cases}$$

Show that for any interval (a, b) in $[0, 1)$ one has that $v_G((a, b)) = v_G(T_G^{-1}(a, b))$, where v_G is the probability measure on $[0, 1)$ that satisfies

$$v_G((a, b)) = \int_a^b g_G(x)dx \,. \qquad \blacksquare$$

The existence of such measures v_β for each $\beta > 1$ was shown by A. Rényi [Rén57] in 1957, who obtained the exact form only for $\beta = G$. The measure was explicitly determined for all $\beta > 1$ by A. O. Gelfond [Gel59] in 1959 and independently by W. Parry [Par60] in 1960. In fact, one of the goals of this book is to give a way of obtaining this invariant measure v_β. This is done by connecting the β-expansion with an (appropriate) GLS expansion. The technique used can be extended to a wide class of piecewise linear transformations.

2.4.2 Periodic β-expansions

In the cases of n-ary and continued fraction expansions we saw that eventually-periodic expansions can be easily characterized. At first sight β-expansions might look like trivial variations of n-ary expansions, but their simplicity is deceptive! We already saw in the previous subsection that the map T_β no longer has Lebesgue measure as invariant measure when $\beta \notin \mathbb{N}$. It turns out that it is also more difficult to "understand" eventually-periodic expansions. We will sketch here to some extent what is known and mention a longstanding conjecture. Basic references are [Sch80a] and [Bla89].

We first introduce some notations.

Definition 2.4.3. *Let $\beta > 1$ be an algebraic integer; that is, β is the root of a polynomial with integer coefficients, of which the leading coefficient is 1. Then β is called a Pisot number if all its conjugate roots have modulus strictly smaller than 1. We call β a Salem number if all its conjugates z have modulus $|z| \leq 1$, and if there is at least one conjugate of modulus 1.*

Examples 2.4.4. The smallest Pisot number θ_0 is the real root of $P_0(z) = z^3 - z - 1$, and equals $\theta_0 = 1.3247179572\ldots$. The next Pisot number θ_1 is the positive root of $P_1(z) = z^4 - z^3 - 1$, and equals $\theta_1 = 1.3802775691\ldots$. Clearly the 'golden mean' is also a Pisot number. The set of all Pisot numbers is an infinite set, which follows because all integers are Pisot, but also because A. Brauer [Bra51] showed that for all $m \geq 2$ and all $a_1, \ldots, a_m \in \mathbb{N}$, where $a_1 \geq a_2 \geq \cdots \geq a_m$, one has that the positive root θ of $P(z) = z^m - a_1 z^{m-1} - \cdots - a_m$ is a Pisot number. R. Salem [Sal63] showed in 1945 that every Pisot number is the limit of an increasing sequence and a decreasing sequence of Salem numbers. ∎

Exercise 2.4.5. Show by elementary means that θ_0 and θ_1 are Pisot. ∎

Proposition 2.4.6. *Let $\beta > 1$ and let $\mathrm{Per}(\beta)$ be the set of those $x \in [0, 1)$ for which $\{T_\beta^k x; \ k \geq 0\}$ is finite (such x are also called β-numbers). Furthermore, let $\mathbb{Q}(\beta)$ be the smallest field containing both $\mathbb{Q}$ and β. We have $\mathrm{Per}(\beta) \subset \mathbb{Q}(\beta) \cap [0, 1)$.*

Proof. Let $x \in \mathrm{Per}(\beta)$; then there exist integers $m \geq 1$ and $\ell \geq 1$ such that the β-expansion of x is given by

$$x = \frac{\varepsilon_1}{\beta} + \cdots + \frac{\varepsilon_{m-1}}{\beta^{m-1}} + \frac{\varepsilon_m}{\beta^m} + \cdots + \frac{\varepsilon_{m+\ell-1}}{\beta_{m+\ell-1}} + \cdots ;$$

where $\varepsilon_{k+\ell} = \varepsilon_k$, for $k \geq m$. Abbreviating this symbolically gives

$$x = .\varepsilon_1 \cdots \varepsilon_{m-1} \overline{\varepsilon_m \cdots \varepsilon_{m+\ell-1}} ,$$

where the bar—as in Chapter 1—indicates the period. We now will recycle the idea behind the solution of Exercise 1.1.2. Setting $y = \overline{.\varepsilon_m \cdots \varepsilon_{m+\ell-1}}$ one has

$$y = \frac{\varepsilon_m}{\beta} + \cdots + \frac{\varepsilon_{m+\ell-1}}{\beta^\ell} + \frac{y}{\beta^\ell} .$$

From this it follows that

$$y = \frac{\beta^{\ell-1}}{\beta^\ell - 1}\varepsilon_m + \frac{\beta^{\ell-2}}{\beta^\ell - 1}\varepsilon_{m+1} + \cdots + \frac{\varepsilon_{m+\ell-1}}{\beta^\ell - 1} \in \mathbb{Q}(\beta) \cap [0, 1) ,$$

and therefore

$$x = \frac{\varepsilon_1}{\beta} + \cdots + \frac{\varepsilon_{m-1}}{\beta^{m-1}} + \frac{y}{\beta^{m-1}} \in \mathbb{Q}(\beta) \cap [0, 1) . \qquad \blacksquare$$

It turns out that the converse statement is much harder than it is in the case $\beta \in \mathbb{N}$, $\beta > 1$ (in that case $\mathbb{Q}(\beta) = \mathbb{Q}$ and $\mathrm{Per}(\beta) = \mathbb{Q}$). In 1980 K. Schmidt [Sch80a] obtained the following results, some of which were previously obtained by A. Bertrand [Ber77].

Theorem 2.4.7. (K. Schmidt, 1980) *Let $\beta > 1$ be such that $\mathbb{Q} \cap [0, 1) \subset \mathrm{Per}(\beta)$. Then β is either a Pisot number or a Salem number.*

Theorem 2.4.8. (A. Bertrand, 1977, K. Schmidt, 1980) *Let $\beta > 1$ be a Pisot number. Then*

$$\mathrm{Per}(\beta) = \mathbb{Q}(\beta) \cap [0, 1) .$$

Schmidt also proposed the following conjecture.

Conjecture 2.4.9. (K. Schmidt) *Let $\beta > 1$ be a Salem number. Then*

$$\text{Per}(\beta) = \mathbb{Q}(\beta) \cap [0, 1) \,.$$

This conjecture has been verified by D. Boyd for all Salem numbers of degree 4. Further, a heuristic model given by Boyd in [Boy96], suggests that all or 'almost all' Salem numbers x of degree 6 will be β-numbers (i.e., $x \in \text{Per}(\beta)$), but that, for degree 8 and higher, a positive proportion of Salem numbers will not be eventually-periodic.

CHAPTER **3**
Ergodicity

3.1 The Ergodic Theorem

Why is the existence of a T-invariant measure important? The reason is that it allows us to use ergodic theory to study the behavior of the map T. In particular, it makes available the use of the Ergodic Theorem with which answers to many number-theoretical questions can be given (like frequency of digits).

The word *ergodic* comes from the Greek (*ergon* = work and *odos* = path), and originates from physics. In the 1920s and 1930s it drew the attention of mathematicians. In this chapter we will present some of the basic results in ergodic theory.

3.1.1 Integrals

Before we proceed we give a short introduction to the notion of an *integral* of a measurable function (or, as known in probability theory, the *expectation* of a random variable). Let $(X, \mathcal{F}, \mu)$ be a probability space; the definition of an integral is now given in four steps.

(I) First, let $B \in \mathcal{F}$ and consider I_B, the *indicator function* of B, defined by

$$I_B(x) := \begin{cases} 1, & \text{if } x \in B, \\ 0, & \text{if } x \notin B. \end{cases}$$

Define the integral of I_B by $\int_X I_B(x)d\mu = \mu(B)$.

(II) Now let $B_1, \ldots, B_n \in \mathcal{F}$ be pairwise disjoint, and consider the *simple* function $\phi(x) = \sum_{i=1}^n a_i I_{B_i}(x)$, so that $\phi(x) = a_i$ if $x \in B_i$. We define

$$\int_X \phi(x)d\mu = \sum_{i=1}^n a_i \mu(B_i) .$$

If $C \in \mathcal{F}$ is any measurable set, we define

$$\int_C \phi(x)d\mu = \int_X \phi(x)I_C(x)d\mu = \sum_{i=1}^n a_i \mu(B_i \cap C) .$$

(III) If $f : X \to [0, \infty)$ is $\mathcal{B}$-measurable, then we define

$$\int_X f(x)d\mu := \sup\left(\int_X \phi(x)d\mu : \phi \text{ is simple and } 0 \le \phi \le f \right) .$$

(IV) If $h : X \to \mathbb{R}$ is $\mathcal{B}$-measurable, we can write $h(x) = h^+(x) - h^-(x)$, where $h^+(x) := \max(0, h(x))$ and $h^-(x) := \max(0, -h(x))$ for every $x \in X$. Then $h^+, h^- \ge 0$, and in case both $\int_X h^+(x)d\mu < \infty$ and $\int_X h^-(x)d\mu < \infty$ we define $\int_X h(x)d\mu$ by

$$\int_X h^+(x)d\mu - \int_X h^-(x)d\mu .$$

We call h *integrable* if h is $\mathcal{B}$-measurable and $\int_X |h(x)|d\mu < \infty$. In this case $\int_X h(x)d\mu$ is finite; we write $h \in L^1(X, \mathcal{F}, \mu)$.

We say that a property holds *almost everywhere* (in short, a.e.), if it holds for all points x outside a set of measure zero.

In case $(\Omega, \mathcal{F}, P)$ is a probability space, i.e., $P(\Omega) = 1$ and Z is a random variable defined on Ω, then the *expectation* $E(Z)$ of Z is $\int_\Omega Z(\omega)dP$, whenever Z is integrable.

Exercise 3.1.1. Let $(X, \mathcal{F}, \mu)$ be a measure space, and let $f : X \to [0, \infty)$ be integrable. Define $\rho : \mathcal{F} \to \mathbb{R}$ by

$$\rho(A) := \int_A f(x) d\mu.$$

(a) Show that ρ defines a measure on $(X, \mathcal{F})$.

(b) Show that if $\mu(A) = 0$, then $\rho(A) = 0$. Furthermore, if $f > 0$ a.e., then $\rho(A) = 0$ implies $\mu(A) = 0$. So we see that ρ and μ are *equivalent*. ∎

Exercise 3.1.2. Using the same notation as in Exercise 3.1.1, show that

$$\int_X f d\mu = 0 \quad \text{implies that} \quad f = 0 \ \mu \text{-a.e.} \qquad ∎$$

For more information on integrals and measure theory we refer to the excellent books by W. Rudin [Rud87] and H.L. Royden [Roy88].

3.1.2   The Ergodic Theorem

The Ergodic Theorem is also known as Birkhoff's Ergodic Theorem or the Individual Ergodic Theorem (1931). This theorem is in fact a generalization of the strong law of large numbers (SLLN); see [KT66]. Let $X_1, X_2, \ldots$ be i.i.d. random variables on a probability space $(X, \mathcal{F}, \nu)$, with $E|X_i| < \infty$; then

$$\lim_{n \to \infty} \frac{1}{n} \sum_{i=1}^{n} X_i = EX_1 \ \text{(a.e.)}.$$

For the GLS family the SLLN is in fact enough to investigate the distribution of the digits. As an example we will show here that almost every x (with respect to Lebesgue measure λ) is *simple normal*. Let us first define what simple normal numbers are.

Let $\mathcal{I} = \{[\ell_n, r_n) : n \in \mathcal{D}\}$ be the partition of the GLS($\mathcal{I}$) transformation under consideration, and suppose $x \in [0, 1)$ has GLS($\mathcal{I}$)

expansion

$$x = \sum_{k=1}^{\infty} \frac{h_k}{s_1 s_2 \cdots s_k}. \tag{3.1}$$

Let $x = [a_1, a_2, \ldots, a_k, \ldots]$ be the sequence of digits of x corresponding to the GLS($\mathcal{I}$) expansion (3.1). We call x a *simple normal number* for the GLS($\mathcal{I}$) expansion if, for all $j \in \mathcal{D}$,

$$\lim_{n \to \infty} \frac{1}{n} \#\{1 \le i \le n : a_i = a_i(x) = j\}$$
$$= \lambda\{y : a_1(y) = j\} = L_j = r_j - \ell_j,$$

where $\#A$ is the *cardinality* of the set A, i.e., the number of elements in A. Fix $j \in \mathcal{D}$ and let $X_1, X_2, \ldots$ be a sequence of random variables on $([0, 1), \lambda)$, defined as follows:

$$X_i(y) = \begin{cases} 1 & \text{if } a_i(y) = j, \\ 0 & \text{if } a_i(y) \neq j. \end{cases}$$

Then $X_1, X_2, \ldots$ are i.i.d. (since the digits are independent and identically distributed) and for each n

$$\sum_{i=1}^{n} X_i(y) = \#\{1 \le i \le n : a_i(y) = j\}.$$

But then it follows from the SLLN that for a.e. y,

$$\lim_{n \to \infty} \frac{1}{n} \#\{1 \le i \le n : a_i(y) = j\} = \lim_{n \to \infty} \frac{1}{n} \sum_{i=1}^{n} X_i(y)$$
$$= EX_1 = \lambda(a_1(y) = j).$$

This is known as *Borel's Normal Number Theorem*.

Definition 3.1.3. *A number $x \in [0, 1)$ with GLS($\mathcal{I}$) expansion (3.1) and sequence of digits $x = [a_1, a_2, \ldots, a_k, \ldots]$ is said to* *be normal if, for any positive integer m and any block of digits* $b_0 b_1 \ldots b_{m-1}$ $(b_i \in \mathcal{D})$,

$$\lim_{n \to \infty} \frac{1}{n} \#\{1 \le i \le n - m : a_i = b_0, a_{i+1} = b_1, \ldots, a_{i+m-1} = b_{m-1}\}$$

$$= \lambda\{y : a_1(y) = b_0, a_2(y) = b_1, \ldots, a_m(y) = b_{m-1}\}$$

$$= L_{b_0} L_{b_1} \ldots L_{b_{m-1}}.$$

Here the strong law of large numbers cannot be used directly to check normality. To see this, consider for example the binary expansion. In this case $\mathcal{I} = \left\{[0, \frac{1}{2}), [\frac{1}{2}, 1)\right\} = \{I_0, I_1\}$, and digit set $\mathcal{D} = \{0, 1\}$. Suppose we want to study the frequency of the block 01. Write $y \in [0, 1)$ in terms of its binary expansion $y = [y_1, y_2, y_3, \ldots]$, $y_i \in \mathcal{D}$. Define the sequence of random variables $X_1, X_2, \ldots$ as follows:

$$X_i(y) := \begin{cases} 1, & \text{if } y_i y_{i+1} = 01, \\ 0, & \text{otherwise.} \end{cases}$$

Then

$$\sum_{i=1}^n X_i(y) = \#\{1 \le i \le n : y_i y_{i+1} = 01\}$$

$$= \text{number of times 01 occurred in } y_1 \ldots y_{n+1}.$$

The binary map $Tx = 2x \pmod 1$ is measure preserving with $\lambda(X_1 = 1) = \lambda(X_i = 1)$, for all i, hence $X_1, X_2, \ldots$ are identically distributed. However, $X_1, X_2, \ldots$ are *not* independent, since for example

$$\lambda(X_1 = 1, X_2 = 0)$$

$$= \lambda(\{y = [y_1, y_2, \ldots] : y_1 y_2 = 01 \text{ and } y_2 y_3 \ne 01\})$$

$$= \lambda(X_1 = 1) = \frac{1}{4}$$

$$\ne \lambda(X_1 = 1)\lambda(X_2 = 0) = \frac{1}{16}.$$

How can we study the normality of numbers? One way to do this is to replace the SLLN by a more general theorem, known as the Ergodic

Theorem, which requires a weaker notion of independence known as *ergodicity*. Heuristically, a dynamical system $(X, \mathcal{F}, \mu, T)$ is ergodic if it cannot be seen as the union of two separate dynamical systems.

Definition 3.1.4. *Let $(X, \mathcal{F}, \mu, T)$ be a dynamical system. Then T is called ergodic if for every μ-measurable set A satisfying $T^{-1}A = A$ (such a set is called T-invariant) one has that $\mu(A) = 0$ or 1.*

In case T is invertible, the above definition is equivalent to: T is ergodic if and only if $A = TA$ implies $\mu(A) = 0$ or 1.

There are many handy characterizations of ergodicity; here we list some of them. For the proof of the following Proposition, see [Wal82], Theorem 1.5. Another characterization will be given in Proposition 3.1.9.

Proposition 3.1.5. (Characterization of ergodicity) *Let $(X, \mathcal{F}, \mu, T)$ be a dynamical system. Then the following statements are equivalent.*

(i) *T is ergodic.*

(ii) *For every $A \in \mathcal{F}$ with $\mu(T^{-1}A \triangle A) = 0$, one has that $\mu(A) = 0$ or 1.*

(iii) *For every $A \in \mathcal{F}$ of positive measure, one has that*

$$\mu \left(\bigcup_{n=1}^{\infty} T^{-n} A \right) = 1.$$

(iv) *For every $A, B \in \mathcal{F}$ of positive measures, there exists a positive integer n such that $\mu(T^{-n} A \cap B) > 0$.*

Remarks 3.1.6.

1. In case T is invertible, then in the above characterization one can replace T^{-n} by T^{n}.

2. Note that if $\mu(A \triangle T^{-1}A) = 0$, then $\mu(A \backslash T^{-1}A) = \mu(T^{-1}A \backslash A) = 0$. Since

$$A = \left(A \setminus T^{-1}A \right) \cup \left(A \cap T^{-1}A \right),$$

and

$$T^{-1}A = \left(T^{-1}A \setminus A\right) \cup \left(A \cap T^{-1}A\right),$$

we see that after removing a set of measure 0 from A and a set of measure 0 from $T^{-1}A$, the remaining parts are equal. In this case we say that A equals $T^{-1}A$ modulo sets of measure 0.

3. In words, (iii) says that if A is a set of positive measure, almost every $x \in X$ eventually (in fact infinitely often) will visit A. ∎

Theorem 3.1.7. (The Ergodic Theorem) *Let $(X, \mathcal{F}, \mu)$ be a probability space and $T : X \rightarrow X$ a measure preserving transformation. Then, for any f in $L^1(\mu)$,*

$$\lim_{n \to \infty} \frac{1}{n} \sum_{i=0}^{n-1} f \circ T^i(x) = f^*(x)$$

exists a.e., is T-invariant and $\int_X f d\mu = \int_X f^* d\mu$. If moreover T is ergodic, then f^* is a constant a.e. and $f^* = \int_X f d\mu$.

Remark 3.1.8. The Ergodic Theorem was originally proved by G.D. Birkhoff in 1931. Since then, several proofs of this important theorem have been obtained; see for instance the books by P. Walters [Wal82] and K. Petersen [Pet89]. Here we present a special case of a recent and rather simple version of the proof by Y. Katznelson and B. Weiss [KW82], initially given by T. Kamae [Kam82] in the setting of non-standard analysis. For the complete proof we refer the reader to the original articles [KW82] or [Kam82], or to the recent book by G. Keller [Kel98]. ∎

Proof. We do the proof only for the case $f = I_B$, the indicator function of some measurable subset B of X. In this case

$$\sum_{i=0}^{n-1} f(T^i x) = \#\{0 \le i \le n - 1 : T^i x \in B\}.$$

Define

$$\overline{f}(x) = \limsup_{n \to \infty} \frac{1}{n} \sum_{i=0}^{n-1} I_B \circ T^i(x) \qquad \text{and}$$

$$\underline{f}(x) = \liminf_{n \to \infty} \frac{1}{n} \sum_{i=0}^{n-1} I_B \circ T^i(x).$$

Note that both $\overline{f}$ and $\underline{f}$ exist, are measurable (see Exercise 1.2.10) and $0 \le \underline{f}(x) \le \overline{f}(x) \le 1$ for all $x \in X$. Also $\overline{f}$ is T-invariant since

$$\overline{f}(Tx) = \limsup_{n \to \infty} \left(\frac{n+1}{n} \frac{1}{n+1} \sum_{i=0}^{n} I_B \circ T^i(x) - \frac{I_B(x)}{n} \right) = \overline{f}(x).$$

A similar argument shows that $\underline{f}$ is T-invariant. To prove the Ergodic Theorem it is enough to show that

$$\int_X \overline{f} d\mu \le \int_X f d\mu = \int_X I_B d\mu = \mu(B) \le \int_X \underline{f} d\mu.$$

For then this implies that

$$\int_X \overline{f} d\mu = \mu(B) = \int_X \underline{f} d\mu.$$

Since $\overline{f} - \underline{f} \ge 0$ it follows by Exercise 3.1.2 that $\overline{f} = \underline{f}$ a.e. We call their common value f^*, and the result follows.

We first show that $\int \overline{f} d\mu \le \mu(B)$. Let $\epsilon > 0$ be given and let $S_n(x) = \sum_{i=0}^{n-1} I_B \circ T^i(x)$. For each $x \in X$, there exist infinitely many integers $n \ge 1$ such that $S_n(x) \ge (\overline{f}(x) - \epsilon)n$. Let

$$N(x) = \min\{n \ge 1 : S_n(x) \ge (\overline{f}(x) - \epsilon)n\}.$$

In particular,

$$S_{N(x)}(x) \ge (\overline{f}(x) - \epsilon)N(x).$$

Since $N(x) < \infty$ for all $x \in X$, it follows that there exists $M > 0$ such that

$$\mu(\{x \in X : N(x) > M\}) < \epsilon.$$

Let $B' = B \cup \{x \in X : N(x) > M\}$ and $S'_n(x) = \frac{1}{n} \sum_{i=0}^{n-1} I_{B'} \circ T^i(x)$. Define

$$N'(x) := \begin{cases} N(x), & \text{if } N(x) \leq M, \\ 1, & \text{if } N(x) > M. \end{cases}$$

Note that $N'(x) \leq M$ for all $x \in X$, and

- if $N(x) > M$, then $x \in B'$ so that

$$S'_{N'(x)}(x) = I_{B'}(x) = 1 > \overline{f}(x) - \epsilon = (\overline{f}(x) - \epsilon)N'(x).$$

- if $N(x) \leq M$, then since $B \subset B'$, we have

$$S'_{N'(x)}(x) = \sum_{i=0}^{N(x)-1} I_{B'} \circ T^i(x)$$

$$\geq \sum_{i=0}^{N(x)-1} I_B \circ T^i(x) \geq (\overline{f}(x) - \epsilon)N'(x).$$

From the above we see that $S'_{N'(x)}(x) \geq (\overline{f}(x) - \epsilon)N'(x)$ for all $x \in X$. Define $n_0(x) = 0$ and $n_k(x) = n_{k-1}(x) + N'(T^{n_{k-1}(x)}x)$, for $k \geq 1$. Choose n much bigger than M, and let $l = \max\{k \geq 1 : n_k(x) \leq n - 1\}$. Using the T-invariance of $\overline{f}$ one has,

$$S'_n(x) \geq \sum_{i=0}^{n_l(x)-1} I_{B'} \circ T^i(x) = \sum_{i=0}^{l-1} \sum_{j=n_i(x)}^{n_{i+1}(x)-1} I_{B'} \circ T^j(x)$$

$$= \sum_{i=0}^{l-1} S'_{N'(T^{n_i(x)}x)}(T^{n_i}(x))$$

$$\geq \sum_{i=0}^{\ell-1} N'(T^{n_i(x)}x)(\overline{f}(T^{n_i(x)}x) - \epsilon)$$

$$= \sum_{i=0}^{l-1}(n_{i+1}(x) - n_i(x))(\overline{f}(x) - \epsilon)$$

$$\geq (n - M)(\overline{f}(x) - \epsilon).$$

Dividing by n, integrating and using the fact that T is measure preserving, we get

$$\mu(B') = \frac{1}{n} \int_X S'_n(x) d\mu \geq \frac{n-M}{n} \left(\int_X \overline{f} d\mu - \epsilon \right).$$

Taking limits gives $\mu(B') \geq \int_X \overline{f} d\mu - \epsilon$. However, $\mu(B') \leq \mu(B) + \epsilon$; thus $\mu(B) \geq \int_X \overline{f} d\mu - 2\epsilon$. Since $\epsilon > 0$ was arbitrary, it follows that $\mu(B) \geq \int_X \overline{f} d\mu$, as required.

Next we show that $\int_X \underline{f} d\mu \geq \mu(B)$. To do this we apply the previous argument to $X \setminus B$ as follows. Let $g = I_{X \setminus B} = 1 - I_B$. Then, $\overline{g}(x) = 1 - \underline{f}(x)$. Now,

$$1 - \int_X \underline{f} d\mu = \int_X \overline{g}(x) \mu \leq \mu(X \setminus B) = 1 - \mu(B).$$

That f^* is a constant a.e. if T is ergodic follows from Proposition 3.1.9 below. This completes the proof. ∎

Proposition 3.1.9. *A measure preserving transformation T on a probability space $(X, \mathcal{B}, \mu)$ is ergodic if and only if every T-invariant measurable function f (i.e., $f \circ T = T$ a.e.) is a constant almost everywhere.*

Proof. Suppose every T-invariant function is a constant almost surely. Let A be a T-invariant set and consider the indicator function I_A of A. Since

$$I_A(Tx) = I_{T^{-1}A}(x) = I_A(x),$$

it follows that I_A is a T-invariant function. Notice that I_A attains only the values 0 and 1, so that either I_A is 0 or I_A is 1 almost everywhere. Since $\mu(A) = \int_X I_A d\mu$ it follows that $\mu(A)$ is either 0 or 1, and T is ergodic.

Conversely, suppose that T is ergodic and that f is T-invariant. For each $r \in \mathbb{R}$, define A_r as

$$A_r = \{x \in X; f(x) > r\}.$$

Since f is T-invariant it follows that A_r is a T-invariant set, and so by ergodicity of T we must have that $\mu(A_r)$ is 0 or 1. Now suppose that f is *not* a constant a.e. Then there must exist an $r \in \mathbb{R}$ such that $0 < \mu(A_r) < 1$, which is a contradiction.   ∎

The following exercise is in fact a classical theorem in ergodic theory. We advise you to try to prove it yourself, but in case of difficulty, see [Wal82], p. 41.

Exercise 3.1.10. Let $(X, \mathcal{F}, \mu)$ be a probability space, and let $\mathcal{A}$ be a generating semi-algebra. Let $T : X \to X$ be a measure preserving transformation; then T is ergodic if and only if for every $A, B \in \mathcal{A}$

$$\lim_{n \to \infty} \frac{1}{n} \sum_{i=0}^{n-1} \mu(T^{-i} A \cap B) = \mu(A)\mu(B).$$ ∎

From the above exercise one can interpret ergodicity as a weak form of independence.

Exercise 3.1.11. Use Exercise 3.1.10 to show that every Bernoulli shift is ergodic.   ∎

Exercise 3.1.12. Show with the help of the Ergodic Theorem that if T is ergodic, then for every set B of positive measure and for a.e. $x \in X$ one has $T^n x \in B$ infinitely often.   ∎

In order to apply the Ergodic Theorem to study the frequency of blocks generated by a GLS transformation, we must first prove ergodicity. For this the following lemma is very useful.

Lemma 3.1.13. (Knopp) *If B is a Lebesgue set and $\mathcal{C}$ is a class of subintervals of $[0, 1)$ satisfying*

(a) *every open subinterval of $[0, 1)$ is at most a countable union of disjoint elements from $\mathcal{C}$,*

(b) $\forall A \in \mathcal{C}$, $\lambda(A \cap B) \geq \gamma \lambda(A)$, where $\gamma > 0$ is independent of A,

then $\lambda(B) = 1$.

Proof. The proof is done by contradiction. Suppose $\lambda(B^c) > 0$. Since B is a Lebesgue set, then $B = C \cup D$ with C a Borel set and D a subset of a Borel set of Borel measure zero. Further the Lebesgue measure $\lambda(B)$ of B is equal to the Borel measure $\lambda(C)$ of C (recall that the completion λ^* of λ is also denoted by λ; see the discussion following Exercise 1.2.6), and if $\lambda(B^c) > 0$, then $\lambda(C^c) > 0$. Given $\varepsilon > 0$ there exists by Theorem 1.2.7 a set E_ε that is a finite disjoint union of open intervals such that $\lambda(C^c \triangle E_\varepsilon) < \varepsilon$. Now by conditions (a) and (b) (that is, writing E_ε as a countable union of disjoint elements of $\mathcal{C}$) one gets that $\lambda(C \cap E_\varepsilon) \geq \gamma \lambda(E_\varepsilon)$.

Also from our choice of E_ε and the fact that

$$\lambda(C^c \triangle E_\varepsilon) \geq \lambda(C \cap E_\varepsilon) \geq \gamma \lambda(E_\varepsilon) \geq \gamma \lambda(C^c \cap E_\varepsilon) > \gamma(\lambda(C^c) - \varepsilon),$$

we have that

$$\gamma(\lambda(C^c) - \varepsilon) < \lambda(C^c \triangle E_\varepsilon) < \varepsilon.$$

Hence $\gamma \lambda(C^c) < \varepsilon + \gamma \varepsilon$, and since $\varepsilon > 0$ is arbitrary, we get a contradiction.   ■

Theorem 3.1.14. *Let T be a GLS($\mathcal{I}$) transformation on $[0, 1)$; then T is ergodic.*

Proof. Let B be a T-invariant Lebesgue set of positive measure. According to Definition 3.1.4 we need to show that $\lambda(B) = 1$, which will be derived using Knopp's Lemma. Let $\mathcal{C}$ be the collection of all fundamental intervals. Property (a) from Knopp's Lemma is clearly satisfied. Now, since T has constant slope on each fundamental interval E of rank n, it follows that

$$\frac{\lambda(A \cap E)}{\lambda(E)} = \frac{\lambda(T^{-n}A \cap E)}{\lambda(E)} = \frac{\lambda(A \cap T^n E)}{\lambda(T^n E)} = \lambda(A),$$

which implies that $\lambda(B \cap E) = \lambda(B)\lambda(E)$ for every fundamental interval E.

Applying the above Lemma with $\gamma = \lambda(B)$ we get that $\lambda(B) = 1$, i.e., T is ergodic with respect to λ. ∎

Remark 3.1.15. In Exercise 2.3.5 we saw that every GLS transformation is (isomorphic to) a Bernoulli shift. In Exercise 3.1.11 you gave a proof that every Bernoulli shift is ergodic; this gives another reason why every GLS transformation is ergodic. ∎

Let us return to the binary example discussed after Definition 3.1.3. Here,

$$X_i(y) = \begin{cases} 1, & \text{if } y_i y_{i+1} = 01 \\ 0, & \text{otherwise} \end{cases} = X_1 \circ T^{i-1}(y).$$

We can apply the Ergodic Theorem with $f = X_1$. We then have

$$\lim_{n \to \infty} \frac{1}{n} \sum_{i=1}^{n} X_i(y) = \lim_{n \to \infty} \frac{1}{n} \sum_{i=1}^{n} X_1 \circ T^{i-1}(y)$$

$$= \int X_1 d\lambda = \lambda(\{y : y_1 y_2 = 01\}) = \tfrac{1}{4}.$$

In general, given any GLS($\mathcal{I}$) transformation with digit set $\mathcal{D}$, and any admissible block $b_1, \ldots, b_m$, i.e., this block appears in the GLS expansion of some $y \in (0, 1)$, one can study the frequency of occurrence of $b_1, \ldots, b_m$ by considering the function f, defined as follows. Let $[a_1, a_2, \ldots]$ be the sequence of digits of x corresponding to the GLS($\mathcal{I}$) expansion (with digit set $\mathcal{D}$) of x. Write

$$f(y) = \begin{cases} 1, & \text{if } a_1, \ldots, a_m = b_1, \ldots, b_m, \\ 0, & \text{otherwise}; \end{cases}$$

then

$$f(T^{i-1}y) = \begin{cases} 1, & \text{if } a_i, \ldots, a_{i+m-1} = b_1, \ldots, b_m, \\ 0, & \text{otherwise}, \end{cases}$$

and

$$\frac{1}{n}\sum_{i=0}^{n-1} f(T^{i-1}y) = \frac{1}{n}\#\{1 \le i \le n \,:\, a_i = b_1, \ldots, a_{i+m-1} = b_m\},$$

which implies that

$$\frac{1}{n}\sum_{i=0}^{n-1} f(T^{i-1}y) = L_{b_1} L_{b_2} \cdots L_{b_m} \quad \text{a.e.}$$

The above shows that, for any GLS($\mathcal{I}$) expansion, almost every point is normal.

3.1.3 Mixing

So far we have discussed only two kinds of so-called *mixing properties*, ergodicity and Bernoullicity (i.e., isomorphic to a Bernoulli shift). Loosely speaking, by a mixing property we mean the degree of independence and, seen in this perspective, Bernoullicity is the strongest property possible. In between ergodicity and Bernoullicity are various notions of mixing, of which we will mention three. We will be brief by giving only definitions, since ergodicity (the weakest property) is sufficient for our purposes. However, mixing properties in general play an important role in ergodic theory, and we refer the reader to standard ergodic theory books like [Kel98], [Pet89] or [Wal82].

Throughout this subsection we assume that T is a measure preserving transformation on a probability space $(X, \mathcal{F}, \mu)$.

(i) T is *weakly mixing* if for all $A, B \in \mathcal{F}$

$$\lim_{n\to\infty} \frac{1}{n}\sum_{i=0}^{n-1} \left| \mu(T^{-i}A \cap B) - \mu(A)\mu(B) \right| = 0.$$

There are several equivalent definitions of weakly mixing. In particular, the following definition is useful in Section 6.3.

T is weakly mixing if and only if $T \times T$ is ergodic,

where $T \times T$ is a (measure-preserving) transformation on $(X \times X, \mathcal{F} \times \mathcal{F}, \mu \times \mu)$, defined by $T \times T(x, y) = (T(x), T(y))$.

(ii) T is *strong mixing* if for all $A, B \in \mathcal{F}$

$$\lim_{n \to \infty} \mu(T^{-i} A \cap B) = \mu(A)\mu(B).$$

(iii) We say that T is *weak Bernoulli* if for each $\varepsilon > 0$ there exists a positive integer N, such that, for all $m \geq 1$ and for all $A \in \bigvee_{i=0}^{m} T^{-i} \mathcal{F}$ and $C \in \bigvee_{i=-N-m}^{-N} T^{-i} \mathcal{F}$,

$$|\mu(A \cap C) - \mu(A)\mu(C)| < \varepsilon,$$

where $\bigvee_{k}^{\ell} T^{-i} \mathcal{F}$ denotes the smallest σ-algebra containing $T^{-i} \mathcal{F}$ for $i = k, \ldots, \ell$. Notice that weak Bernoulli means that the future and distant past are approximately independent.

Exercise 3.1.16. Show that Bernoullicity implies strong mixing, that strong mixing implies weakly mixing and, finally, that weakly mixing implies ergodicity. ∎

3.2 Examples of normal numbers

Although the existence of normal numbers has been known for some time (Borel's Normal Number Theorem, which we discussed above, dates back to 1909), it was not until 1933 that a concrete normal number was given (or rather constructed) by D.G. Champernowne. The literature on normal numbers is enormous, and almost merits a book on its own. We only mention here—without proofs—some of the earlier results.

3.2.1 Decimal expansion

$$. 1 \, 2 \, 3 \ldots 9 \, 10 \, 11 \, 12 \, 13 \, 14 \, 15 \, 16 \, 17 \, 18 \, 19 \, 20 \, 21 \ldots$$

See also [Cha33]. A beautiful generalization of Champernowne's result was given by H. Davenport and P. Erdös in [DE52]. Let $f(n)$ be any non-constant polynomial in n, all of whose values for $n = 1, 2, \ldots$ are positive integers. Then the decimal number $0.f(1)f(2)\ldots f(n)\ldots$, where $f(n)$ is written in base 10, is normal.

3.2.2 Binary expansion

$$.0\,1\,00\;01\;10\;11\,000\,001\,010\,100\ldots$$

In general for any $n \geq 2$ one can construct a normal number in base n by concatenating successively (in any order) all blocks of length 1, followed by all blocks of length 2, ... etc. . . . .

3.2.3 Continued fractions

Consider the infinite sequence of rational numbers

$$n_1 = \frac{1}{2},\; n_2 = \frac{1}{3},\; n_3 = \frac{2}{3},\; n_4 = \frac{1}{4},\; n_5 = \frac{2}{4},\; n_6 = \frac{3}{4},\; n_7 = \frac{1}{5},\ldots$$

and concatenate the (finite) continued fraction expansions of $n_1, n_2, \ldots$. This yields a number x with continued fraction expansion

$$x = [\,0;\, 2,\, 3,\, 1,\, 2,\, 4,\, 2,\, 1,\, 3,\ldots\,]$$

which is normal with respect to the continued fraction partition. This construction is due to R. Adler, M.S. Keane and M. Smorodinsky; see also [AKS81]. Another construction was given by A.G. Postnikov [Pos60] in 1960.

Remarks 3.2.1.

1. There are many constructions of normal numbers; see e.g. [SV94], [Wag95], and [BM96], and the references in these papers. In [AD79], Y.N. Dowker and J. Auslander apply Furstenberg's concept of disjointness [Fur67] to construct new normal numbers from a given one.

2. Although the above examples give normal numbers, one can not escape the feeling that these numbers are somehow not really random (e.g., for the Champernowne number one can easily compute any of its digits, without knowing the preceding ones). It is for this reason that other concepts—like Kolmogorov complexity—of randomness have been developed. It is beyond the scope of this book to discuss such subjects, but the interested reader is referred to Paul Vitányi's survey paper [Vit95]. ∎

3.3   β-transformations

Let $\beta > 1$ be real. Recall that the transformation $T_\beta : [0, 1) \to [0, 1)$, given by $T_\beta x = \beta x \pmod 1$, generates a series expansion

$$x = \sum_{k=1}^{\infty} \frac{d_k}{\beta^k}$$

where $d_k = d_k(x) = \lfloor \beta T_\beta^{k-1} x \rfloor$, for $k \geq 1$. We denote the β-expansion of $x \in [0, 1)$ by $.d_1 d_2 \ldots d_n \ldots$, to stress the similarity with n-ary expansion. Although 1 is not in the domain of T_β, one can still speak of the β-expansion of 1, denoted by $d(1, \beta) = .b_1 b_2 \cdots$, where $b_i = \lfloor \beta T_\beta^{i-1} 1 \rfloor$ with $T_\beta 1 = \beta - \lfloor \beta \rfloor$.

Interestingly enough, given a non-integer $\beta > 1$, not every infinite series of the form $\sum d_k / \beta^k$, $d_k \in \{0, 1, \ldots, \lfloor \beta \rfloor\}$ is the β-expansion of some number x.

Definition 3.3.1. *We say that the sequence $(d_1, d_2, \ldots, d_k, \ldots)$ is admissible if*

$$\sum_{k=1}^{\infty} \frac{d_k}{\beta^k}$$

is the β-expansion of a point $x \in [0, 1)$.

Exercise 3.3.2. The β-expansion is sometimes known as the *greedy algorithm*; see also [EJK90]. The so-called *lazy expansion* generates expansions of each $x \in [0, 1)$ of the above form that are not β-expansions. What would be an appropriate definition of such *lazy* expansions? ∎

The fact that for a β-expansion not all sequences of digits are admissible is one of the most important reasons why β-expansions are studied. In general, if $\mathcal{A}_n$ is a finite set of n symbols, say,

$$\mathcal{A}_n = \{0, 1, 2, \ldots, n - 1\},$$

and if we look at all one- (or two-) sided sequences of elements from $\mathcal{A}_n$, then the admissible sequences of the β-expansions for a non-integer $\beta > 1$ form a proper closed and shift-invariant subset $\mathcal{D}_\beta$ of $\mathcal{A}_n^{\mathbb{N}}$, where $n = \lfloor \beta \rfloor$. For $\beta = \frac{1}{2}(1 + \sqrt{5})$ we saw in Exercise 2.4.1 that a sequence $d_1 d_2 \ldots d_k \ldots$ of 0's and 1's is an admissible sequence if and only if one never has that two consecutive d_i's equal 11. So one can characterize the set of admissible sequences $\mathcal{D}_\beta$ by giving a finite set of *forbidden words* (in general, a *word* is simply a finite string of symbols from the "alphabet" $\mathcal{A}_n$). This holds for some non-integer $\beta > 1$, and one speaks of *subshifts of finite type*. These subshifts of finite type have many applications, e.g., in coding theory, transmission and storage of data or tilings. For further reading the books of D. Lind and B. Marcus [LM95] and B.P. Kitchens [Kit98] are excellent.

The following proposition, due to W. Parry [Par60], gives a characterization of all admissible sequences (i.e., all sequences that give rise to β-expansions). In a way it explains the nickname of the β-expansion: *greedy algorithm*. We state this proposition without proof.

Proposition 3.3.3. *Let* $\mathbb{W} := \{0, 1, 2, \ldots, \lfloor \beta \rfloor\}^{\mathbb{N}}$ *and let* $\mathcal{D}_\beta$ *be the set of admissible sequences obtained from the β-expansions of real numbers* $x \in [0, 1)$. *Furthermore, let* σ *be the shift and* $<_{\text{lex}}$ *be the lexicographical ordering on* $\mathbb{W}$.

(i) *When the β-expansion $d(1, \beta)$ of 1 is not finite, the condition*

$$\sigma^n(s) <_{\text{lex}} d(1, \beta) \ , \ \text{for all } n \geq 0$$

is necessary and sufficient for the sequence $s \in$ W to belong to $\mathcal{D}_\beta$.

(ii) *If $d(1, \beta) = .b_1b_2\ldots b_k 0000\ldots$, that is, the β-expansion $d(1, \beta)$ of 1 is finite, then $s \in$ W belongs to $\mathcal{D}_\beta$ if and only if for all $n \geq 0$, $\sigma^n(s)$ is lexicographically less than the purely-periodic sequence*

$$d^*(\beta) = .b_1b_2\ldots b_{k-1}(b_k - 1)b_1b_2\ldots b_{k-1}(b_k - 1)b_1b_2$$
$$\ldots b_{k-1}(b_k - 1)\ldots .$$

In case the β-expansion of 1 is infinite, we write $d^*(\beta)$ instead of $d(1, \beta)$.

Examples 3.3.4.

1. Let β be the positive root of the polynomial $z^2 - z - 1 = 0$, i.e., $\beta = \frac{1}{2}(1 + \sqrt{5})$ (the golden mean); then $\beta^2 = 1 + \beta$ and dividing by β^2 gives

$$1 = \frac{1}{\beta} + \frac{1}{\beta^2} \ ,$$

so that 1 has a finite $\beta = \frac{1}{2}(1 + \sqrt{5})$-expansion. Thus, the only sequences that correspond to β-expansions are those that are lexicographically less than $(1010101010\ldots)$. As an example, we see that $(1001001001001\ldots)$ gives rise to a β-expansion while $(11d_3d_4\ldots)$ with $d_i \in \{0, 1\}$ does not.

2. Similarly, consider $\beta > 1$ such that β is the (only) positive root of $z^m - z^{m-1} - \cdots - z - 1 = 0$. Then $\beta^m = \beta^{m-1} + \cdots + \beta + 1$, and dividing by β^m yields

$$1 = \frac{1}{\beta} + \frac{1}{\beta^2} + \cdots + \frac{1}{\beta^m} \ .$$

The only sequences of 0's and 1's that give rise to β-expansions are those that have at most $m - 1$ consecutive 1's. We call such a β a *pseudo-golden mean* of order m.

3. When $\beta > 1$ is an integer, Parry's proposition is trivially true. For if $\beta = n$, then 1 in base n has as expansion

$$1 = (n - 1, n - 1, \ldots, n - 1, \ldots);$$

e.g. in base $n = 10$ one has $1 = (9, 9, 9, \ldots, 9, \ldots)$. Since the digits allowed are elements of $\{0, 1, \ldots, n - 1\}$ it follows that any sequence with digits from $\{0, 1, \ldots, n - 1\}$ not ending in an infinite string of $n - 1$'s gives rise to an n-ary expansion. Conversely, any n-ary expansion yields a sequence lexicographically less than $(n - 1, n - 1, \ldots, n - 1, \ldots)$. ∎

3.4   Ergodic properties of the β-expansion

We consider here the case $\beta > 1$, $\beta \notin \mathbb{Z}$. Rényi [Rén57] showed that

1. T_β is ergodic (we will return to this in a moment).
2. There exists a unique probability measure ν_β, equivalent to the Lebesgue measure λ, which is invariant under T_β. Moreover

$$\nu_\beta(B) = \int_B h_\beta(x)dx,$$

where $h_\beta(x)$ is a measurable function satisfying

$$1 - \frac{1}{\beta} \leq h_\beta(x) \leq \frac{1}{1 - \frac{1}{\beta}}.$$

Shortly afterwards Gelfond [Gel59] and (independently) Parry [Par60] found the invariant measure explicitly, viz.

$$h_\beta(x) = \frac{1}{F(\beta)} \sum_{x < T_\beta^n 1} \frac{1}{\beta^n}, \quad \text{for } x \in [0, 1), \tag{3.2}$$

where the sum runs over all $n \geq 0$ such that $x < T_\beta^n 1$, and

$$F(\beta) = \int_0^1 \left(\sum_{x < T_\beta^n 1} \frac{1}{\beta^n} \right) dx$$

is the normalizing constant.

In the next chapter we will derive formula (3.2), the density of the invariant measure, by connecting any β-expansion with a suitable GLS expansion. First, let us look at an easy but instructive example to stress the importance of the existence of an invariant measure.

Example 3.4.1. Let β be the golden mean, i.e., $\beta = \frac{1}{2}(1 + \sqrt{5})$ which satisfies $\beta^2 - \beta - 1 = 0$. In this case $\frac{1}{\beta} = \frac{1}{2}(-1 + \sqrt{5}) = .6180\cdots$ satisfies $z^2 + z - 1 = 0$. The associated transformation T_β has the following form; see also Figure 3.1.

$$T_\beta x = \begin{cases} \beta x, & 0 \leq x < \frac{1}{\beta}, \\ \beta x - 1, & \frac{1}{\beta} \leq x < 1. \end{cases}$$

Note that for each $0 \leq x < 1/\beta = .6180\ldots$ one has that $x < 1 = T_\beta^0 1$ and $x < 1/\beta = T_\beta^1 1$, while for each $1/\beta \leq x < 1$ one has $x < 1 = T_\beta^0 1$. Since $T_\beta^n 1 = 0$ for $n \geq 2$, a simple calculation yields that $F(\beta) = \frac{1}{2}(5 - \sqrt{5})$ and the density of the invariant measure is given by

$$h_\beta(x) = \begin{cases} \dfrac{5 + 3\sqrt{5}}{10}, & 0 \leq x < \dfrac{\sqrt{5} - 1}{2}, \\ \dfrac{5 + \sqrt{5}}{10}, & \dfrac{\sqrt{5} - 1}{2} \leq x < 1; \end{cases}$$

see also Exercise 2.4.2.

With the invariant measure at hand (and assuming ergodicity of T_β!) we can use the Ergodic Theorem to calculate frequencies of appearances of any given block of digits. We identify points $x \in [0, 1)$ with their associated admissible sequence of 0's and 1's—say $x =$

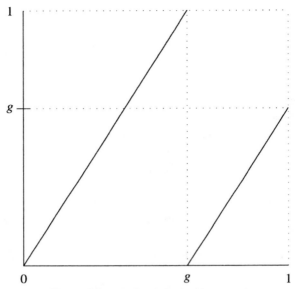

Figure 3.1. T_β for β the 'golden mean'

$(d_1, d_2, \dots)$ where $d_i = d_i(x) \in \{0, 1\}$. Then T_β is just the shift, and for λ (or ν_β) a.e. $x \in [0, 1)$ we have

$$\lim_{n \to \infty} \frac{1}{n} \#\{1 \le i \le n \,:\, d_i(x) = 0\}$$

$$= \lim_{n \to \infty} \frac{1}{n} \sum_{i=0}^{n-1} 1_{[0, 1/\beta)}(T_\beta^i x)$$

$$= \nu_\beta([0, \tfrac{1}{\beta}))$$

$$= \int_0^{1/\beta} \frac{5 + 3\sqrt{5}}{10} dx = \frac{5 + \sqrt{5}}{10} = .7236\dots,$$

which is the frequency of 0's. Similarly for any other block.

It is easy to see that in this case we have

$$1 = \frac{1}{\beta} + \frac{1}{\beta^2} = \frac{1}{\beta} + \frac{1}{\beta^3} + \frac{1}{\beta^5} + \frac{1}{\beta^7} + \cdots,$$

and therefore it follows from Parry's proposition that the only admissible sequences of digits are those in which the block 11 *never* occurs. That is, every 1 is preceded (unless it is at the beginning of a sequence) and followed (unless it is at the end of a finite sequence) by a 0.

Given an admissible block $i_1, i_2, \ldots, i_n$, we denote by

$$\Delta(i_1, i_2, \ldots, i_n) = \{x \,:\, d_1(x) = i_1, \ldots, d_n(x) = i_n\},$$

where

$$x = \sum_{k=1}^{\infty} \frac{d_k(x)}{\beta^k}$$

is the β-expansion of x. We call $\Delta(i_1, i_2, \ldots, i_n)$ a *full interval of rank n* if $\lambda(T_\beta^n \Delta(i_1, i_2, \ldots, i_n)) = 1$, and non-full otherwise. One can easily see that for any full interval Δ_n of rank n one has that $\lambda(\Delta_n) = 1/\beta^n$, since T_β^n on Δ_n is a linear map with slope β^n, attaining the value 0 at the left endpoint of Δ_n, and 1 at the other endpoint. Hence one has that

$$\lambda(\Delta_n) = \frac{1}{\text{slope}} = \frac{1}{\beta^n}\,.$$

Also, notice that the non-full intervals of rank n are precisely those whose last digit i_n equals 1 and that the length of any such interval equals $1/\beta^{n+1}$.

Let $S(n)$ be the number of cylinders of rank n (= all full + non-full intervals of rank n). Now $S(n) = S(n-1) + S(n-2)$ and notice that $S(n-1)$ equals the number of full intervals of rank n and $S(n-2)$ is the number of non-full intervals of rank n. Further, each full interval of rank n has length $1/\beta^n$, while each non-full interval of rank n has length $1/\beta^{n+1}$. Therefore,

$$\frac{S(n)}{\beta^{n+1}} = \frac{S(n-1)}{\beta^{n+1}} + \frac{S(n-2)}{\beta^{n+1}} \leq \frac{S(n-1)}{\beta^n} + \frac{S(n-2)}{\beta^{n+1}} = 1,$$

which implies $S(n) \leq \beta^{n+1}$. ∎

Remark 3.4.2. For $\beta = G = \frac{1}{2}(\sqrt{5} + 1)$ we see that we can cover $[0, 1)$ with disjoint full intervals of rank n or $n + 1$.   ∎

In general one has the following lemma, which we state without proof.

Lemma 3.4.3. *Given any $n \geq 1$ we can cover $[0, 1)$ with disjoint full intervals of rank n or $n + 1$.*

Corollary 3.4.4. *Any interval is at most a countable union of full intervals.*

Theorem 3.4.5. *Let $\beta > 1$. The map $T_\beta x = \beta x \pmod{1}$ is ergodic.*

Proof. Let B be a T_β-invariant Lebesgue set of positive measure. According to Definition 3.1.4 we need to show that $\lambda(B) = 1$, which will be derived using Knopp's Lemma. Let $\mathcal{C}$ be the collection of all full intervals. By Corollary 3.4.4 property (a) of Knopp's Lemma 3.1.13 is clearly satisfied. Now let E be such a full interval of rank n; then for any Lebesgue measurable set C,

$$\lambda(T_\beta^{-n} C \cap E) = \beta^{-n} \lambda(C).$$

Hence,

$$\frac{\lambda(B \cap E)}{\lambda(E)} = \frac{\lambda(T_\beta^{-n} B \cap E)}{\lambda(E)} = \frac{\beta^{-n} \lambda(B)}{\beta^{-n}} = \lambda(B),$$

which implies that $\lambda(B \cap E) = \lambda(B)\lambda(E)$ for every full interval E of rank n. Applying Knopp's Lemma with $\gamma = \lambda(B)$ we get that $\lambda(B) = 1$; i.e., T_β is ergodic with respect to λ.   ∎

3.5 Ergodic properties of continued fractions

In this section we will show that the continued fraction map T is ergodic. To be more precise, we have the following theorem.

Theorem 3.5.1. *Let $\Omega = [0, 1)$, $\mathcal{L}$ the collection of Lebesgue sets of $[0, 1)$ and μ the Gauss-measure on $(\Omega, \mathcal{L})$. Then the dynamical system*

$$(\Omega, \mathcal{L}, \mu, T)$$

is an ergodic system.

Proof. We already saw that T is measure preserving on intervals. Due to Remark 1.2.13 it follows that T is measure preserving.

We now will show that T is ergodic. We have that

$$x = M_n(T^n x),$$

where M_n is the Möbius transformation corresponding to x; see Exercise 1.3.10. One easily shows for a and b, with $0 \le a < b \le 1$, that $\{x : a \le T^n x \le b\} \cap \Delta_n$ equals

$$\left[\frac{p_{n-1}a + p_n}{q_{n-1}a + q_n}, \frac{p_{n-1}b + p_n}{q_{n-1}b + q_n} \right)$$

when n is even, and equals

$$\left(\frac{p_{n-1}b + p_n}{q_{n-1}b + q_n}, \frac{p_{n-1}a + p_n}{q_{n-1}a + q_n} \right]$$

for n odd. Here $\Delta_n = \Delta_n(a_1, \dots, a_n)$ is a fundamental interval of rank n; see also Exercise 1.3.15. Notice that $T^{-n}[a, b) = \{x : a \le T^n x < b\}$.

Exercise 3.5.2. Show that the Lebesgue measure of $T^{-n}[a, b) \cap \Delta_n$ is given by

$$\lambda([a, b)) \lambda(\Delta_n) \frac{q_n(q_{n-1} + q_n)}{(q_{n-1}b + q_n)(q_{n-1}a + q_n)}. \qquad \blacksquare$$

Notice that

$$\frac{1}{2} < \frac{q_n}{q_{n-1} + q_n} < \frac{q_n(q_{n-1} + q_n)}{(q_{n-1}b + q_n)(q_{n-1}a + q_n)}$$

$$< \frac{q_n(q_{n-1} + q_n)}{q_n^2} < 2.$$

Therefore we find for every interval I, that

$$\frac{1}{2}\lambda(I)\lambda(\Delta_n) < \lambda(T^{-n}I \cap \Delta_n) < 2\lambda(I)\lambda(\Delta_n).$$

Let A be a finite disjoint union of such intervals I. Since Lebesgue measure is additive one has

$$\frac{1}{2}\lambda(A)\lambda(\Delta_n) \leq \lambda(T^{-n}A \cap \Delta_n) \leq 2\lambda(A)\lambda(\Delta_n). \qquad (3.3)$$

The collection of finite disjoint unions of such intervals generates the Borel σ-algebra. It follows that (3.3) holds for any Borel set and hence for any Lebesgue set A.

For any $A \in \mathcal{L}$ one has

$$\frac{1}{2\log 2}\lambda(A) \leq \mu(A) \leq \frac{1}{\log 2}\lambda(A); \qquad (3.4)$$

see also Figure 3.2, where the densities of λ and μ are compared. In the above and throughout the book, all logarithms considered are natural logarithms, unless otherwise stated.

Due to (3.3) and (3.4) one has

$$\mu(T^{-n}A \cap \Delta_n) \geq \frac{\log 2}{4}\mu(A)\mu(\Delta_n). \qquad (3.5)$$

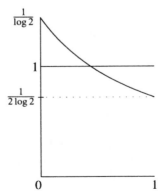

Figure 3.2. The densities of λ and μ

Now let C be the collection of all fundamental intervals Δ_n. Since the set of all endpoints of these fundamental intervals is the set of all rationals in $[0, 1)$, it follows that condition (a) of Knopp's Lemma 3.1.13 is satisfied.

Now suppose that B is invariant with respect to T and $\mu(B) > 0$. Then it follows from (3.5) that for every fundamental interval Δ_n

$$\mu(B \cap \Delta_n) \geq \frac{\log 2}{4} \mu(B) \mu(\Delta_n) .$$

So condition (b) from Knopp's Lemma is satisfied with $\gamma = \frac{\log 2}{4} \mu(B)$; thus $\mu(B) = 1$; i.e., T is ergodic. ∎

Exercise 3.5.3. Show that the continued fraction map T is not Bernoulli, but is strong mixing. ∎

We conclude this chapter with some classical results by Lévy and Khintchine, which can also be found in standard textbooks on continued fractions, like A.Ya. Khintchine's book [Khi63], or the more recent book by A.M. Rockett and P. Szüsz [RS92]. Originally these results were obtained via probability theory; in both cases the so-called *Gauss-Kusmin-Lévy Theorem* is fundamental. This theorem originated from a letter from Gauss to Laplace, dated January 30, 1812. In this letter Gauss stated that he could show that (in modern notation)

$$\lim_{n \to \infty} \lambda(T^{-n}[0, z]) = \mu([0, z]) \quad 0 \leq z \leq 1, \tag{3.6}$$

and asked Laplace to give an estimate of the error term

$$r_n(z) := \lambda(T^{-n}[0, z]) - \mu([0, z]) \quad 0 \leq z \leq 1, n \geq 1.$$

It took more than a century before a proof of (3.6) was published and Gauss' question was answered. In 1928 R. Kusmin [Kus28] showed that

$$r_n(z) = \mathcal{O}(q^{\sqrt{n}}), \quad n \to \infty,$$

where q is some constant, $0 < q < 1$. By this notation we mean that

$$|r_n(z)| \leq Cq^{\sqrt{n}}, \quad \text{as } n \to \infty,$$

for some constant C. Independently and with a different approach using ideas from probability theory, P. Lévy [L29] proved in 1929 that

$$r_n(z) = \mathcal{O}(q^n), \quad n \to \infty,$$

with $q = 0.7\ldots$. Notice that both results are very strong; not only do they give the limit, they also show that the rate of convergence to the limit is exponential. Results like those of Kusmin and Lévy are nowadays known as *Gauss-Kusmin* or *Gauss-Kusmin-Lévy Theorems*.

After Kusmin and Lévy, several improvements (i.e., smaller constants q) of the Gauss-Kusmin Theorem were obtained, see e.g. P. Szüsz [Szü61], F. Schweiger [Sch68], [Sch70], and recently M. Iosifescu [Ios92], [Ios94], [Ios95], [Ios97]. To some extent one could say that the problem found its final solution in the hands of E. Wirsing [Wir74] and K.I. Babenko [Bab78]. For more references and details, see the excellent book of Schweiger [Sch95], and the recent book by Iosifescu and Kraaikamp [IK2002].

The probabilistic theory of continued fractions started with the papers by Kusmin and in particular Lévy; several results (two of which we are about to mention) were obtained using the Gauss-Kusmin Theorem and probability theory, and in general the proofs of these results were by no means obvious. In 1940 Lévy's student W. Doeblin [Doe40] and in 1951 C. Ryll-Nardzewski [RN51] discovered that there is an ergodic system underlying the (regular) continued fraction expansion. In particular, Ryll-Nardzewski showed how several results by Lévy and Khintchine can be obtained using ergodic theory.

Although ergodic theoretic proofs of these results by Lévy and Khintchine can be found in several books, we decided to include them simply because they illustrate the elegance of the use of the Ergodic Theorem. We present some of these results as exercises, the more involved ones as propositions.

Exercise 3.5.4. (Paul Lévy, 1929) Let $a \in \mathbb{N}_+$ be given and let $x \in [0, 1)$, with continued fraction expansion (1.6). Then for almost all $x \in [0, 1)$ one has

$$\lim_{n \to \infty} \frac{1}{n} \#\{1 \le i \le n; \ a_i = a\} = \frac{1}{\log 2} \log \left(1 + \frac{1}{a(a+2)} \right). \quad \blacksquare$$

Thus one sees that for almost every x about $41\frac{1}{2}\%$ of the partial quotients are equal to 1, and slightly fewer than 17% are equal to 2.

Proposition 3.5.5. (Paul Lévy, 1929) *For almost all $x \in [0, 1)$ one has*

$$\lim_{n \to \infty} \frac{1}{n} \log q_n = \frac{\pi^2}{12 \log 2}, \tag{3.7}$$

$$\lim_{n \to \infty} \frac{1}{n} \log(\lambda(\Delta_n)) = \frac{-\pi^2}{6 \log 2}, \quad and \tag{3.8}$$

$$\lim_{n \to \infty} \frac{1}{n} \log \left| x - \frac{p_n}{q_n} \right| = \frac{-\pi^2}{6 \log 2}. \tag{3.9}$$

Proof. By Exercise 1.3.8, for any irrational $x \in [0, 1)$ one has

$$
\frac{1}{q_n(x)} = \frac{1}{q_n(x)} \frac{p_n(x)}{q_{n-1}(Tx)} \frac{p_{n-1}(Tx)}{q_{n-2}(T^2x)} \cdots \frac{p_2(T^{n-2}x)}{q_1(T^{n-1}x)}
$$

$$
= \frac{p_n(x)}{q_n(x)} \frac{p_{n-1}(Tx)}{q_{n-1}(Tx)} \cdots \frac{p_1(T^{n-1}x)}{q_1(T^{n-1}x)}.
$$

Taking logarithms yields

$$
-\log q_n(x) = \log \frac{p_n(x)}{q_n(x)} + \log \frac{p_{n-1}(Tx)}{q_{n-1}(Tx)} + \cdots + \log \frac{p_1(T^{n-1}x)}{q_1(T^{n-1}x)}.
$$
$$\tag{3.10}$$

For any $k \in \mathbb{N}$, and any irrational $x \in [0, 1)$, $\frac{p_k(x)}{q_k(x)}$ is a rational number close to x by (1.14). Therefore we compare the right-hand side of (3.10) with

$$\log x + \log Tx + \log T^2 x + \cdots + \log(T^{n-1}x).$$

We have

$$-\log q_n(x) = \log x + \log Tx + \log T^2 x + \cdots + \log(T^{n-1}x) + R(n, x).$$

In order to estimate the error term $R(n, x)$, we recall from Exercise 1.3.15 that x lies in the interval Δ_n, which has endpoints p_n/q_n and $(p_n + p_{n-1})/(q_n + q_{n-1})$. Therefore, in case n is even, one has

$$0 < \log x - \log \frac{p_n}{q_n} = \left(x - \frac{p_n}{q_n}\right)\frac{1}{\xi} \leq \frac{1}{q_n(q_{n-1} + q_n)}\frac{1}{p_n/q_n} < \frac{1}{q_n},$$

where $\xi \in (\frac{p_n}{q_n}, x)$ is given by the mean value theorem. Let $\mathcal{F}_1, \mathcal{F}_2, \ldots$ be the sequence of Fibonacci $1, 1, 2, 3, 5, \ldots$ (these are the q_i's of the small golden ratio $g = \frac{1}{G}$). It follows from the recurrence relation for the q_i's in Exercise 1.3.8 that $q_n(x) \geq \mathcal{F}_n$. A similar argument shows that

$$\frac{1}{q_n} < \log x - \log \frac{p_n}{q_n},$$

in case n is odd. Thus

$$|R(n, x)| \leq \frac{1}{\mathcal{F}_n} + \frac{1}{\mathcal{F}_{n-1}} + \cdots + \frac{1}{\mathcal{F}_1},$$

and since we have

$$\mathcal{F}_n = \frac{G^n + (-1)^{n+1}g^n}{\sqrt{5}}$$

it follows that $\mathcal{F}_n \sim \frac{1}{\sqrt{5}}G^n$, $n \to \infty$. Thus $\frac{1}{\mathcal{F}_n} + \frac{1}{\mathcal{F}_{n-1}} + \cdots + \frac{1}{\mathcal{F}_1}$ is the nth partial sum of a convergent series, and therefore

$$|R(n, x)| \leq \frac{1}{\mathcal{F}_n} + \cdots + \frac{1}{\mathcal{F}_1} \leq \sum_{n=1}^{\infty} \frac{1}{\mathcal{F}_n} := C.$$

Hence for each x for which

$$\lim_{n \to \infty} \frac{1}{n}(\log x + \log Tx + \log T^2 x + \cdots + \log(T^{n-1}x))$$

exists,

$$- \lim_{n \to \infty} \frac{1}{n} \log q_n(x)$$

exists too, and these limits are equal.

Now $\lim_{n \to \infty} \frac{1}{n}(\log x + \log Tx + \log T^2 x + \cdots + \log(T^{n-1}x))$ is ideally suited for the Ergodic Theorem; we only need to check that the conditions of the Ergodic Theorem are satisfied and to calculate the integral. This is left as an exercise for the reader. This proves (3.7).

It follows from Exercise 1.3.15 that

$$\lambda(\Delta_n(a_1, \ldots, a_n)) = \frac{1}{q_n(q_n + q_{n-1})} \, ;$$

thus

$$- \log 2 - 2 \log q_n < \log \lambda(\Delta_n) < -2 \log q_n \, .$$

Now apply (3.7) to obtain (3.8). Finally (3.9) follows from (3.7) and

$$\frac{1}{2q_n q_{n+1}} < \left| x - \frac{p_n}{q_n} \right| < \frac{1}{q_n q_{n+1}} \, , \ n \geq 1;$$

see also (1.13) in Chapter 1. ∎

Exercise 3.5.6. (A.Ya. Khintchine [Khi35]) For almost every $x \in [0, 1)$ with continued fraction expansion (1.6), one has

$$\lim_{n \to \infty} \frac{n}{\dfrac{1}{a_1} + \cdots + \dfrac{1}{a_n}} = 1.7454056 \ldots. \quad \blacksquare$$

The following exercise is slightly more difficult, the difficulty being the fact that the Ergodic Theorem cannot be applied directly.

Exercise 3.5.7. (A.Ya. Khintchine [Khi35]) For almost every $x \in [0, 1)$ with continued fraction expansion (1.6), one has

$$\lim_{n \to \infty} \frac{a_1 + \cdots + a_n}{n} = \infty. \quad \blacksquare$$

Proposition 3.5.8. (A.Ya. Khintchine [Khi35]) *For almost all x*

$$\lim_{n \to \infty} (a_1 a_2 \cdots a_n)^{1/n} = \prod_{k=1}^{\infty} \left(1 + \frac{1}{k(k+1)} \right)^{\frac{\log k}{\log 2}} = 2.6854 \ldots .$$

Proof. Define for $x \in (0, 1)$ the function $f(x) = \log a_1(x)$, where $a_1(x) = \lfloor \frac{1}{x} \rfloor$; see also Definition 1.3.3. Then $f \in L^1((0, 1), \mathcal{B}, \mu, T)$; i.e., f is an integrable function, since

$$\int_0^1 f \, d\mu = \sum_{k=1}^{\infty} \int_{\frac{1}{k+1}}^{\frac{1}{k}} f \, d\mu \qquad \text{and}$$

$$\int_{\frac{1}{k+1}}^{\frac{1}{k}} f \, d\mu = \frac{1}{\log 2} \int_{\frac{1}{k+1}}^{\frac{1}{k}} \frac{\log a_1(x)}{1+x} \, dx$$

$$= \frac{\log k}{\log 2} \log \left(1 + \frac{1}{k(k+2)} \right) \sim \frac{\log k}{k(k+2)},$$

when $k \to \infty$. Since the series

$$\sum_{k=1}^{\infty} \frac{\log k}{k(k+2)}$$

is convergent it follows that $\int_0^1 f \, d\mu =: a \in \mathbf{R}$.
    But then we have

$$\lim_{n \to \infty} (a_1 a_2 \cdots a_n)^{1/n} = e^a = \prod_{k=1}^{\infty} \left(1 + \frac{1}{k(k+2)} \right)^{\frac{\log k}{\log 2}} = 2.6854 \ldots .$$

∎

CHAPTER **4**

Systems obtained from other systems

In this chapter we will show that there is a deep relationship between GLS expansion and β-expansion. A similar relationship will be used in Chapter 5 to find many metric and Diophantine properties of the regular and (infinitely many) other continued fraction expansions.

4.1 GLS-expansion and β-expansion: A first glimpse at their connection

Let us once more return to the golden mean example. Recall that $\beta = \frac{1}{2}(\sqrt{5} + 1)$ satisfies $z^2 - z - 1 = 0$, and therefore we have

$$1 = \frac{1}{\beta} + \frac{1}{\beta^2} \, ;$$

see also Example 3.4.1. Now T_β is given by

$$T_\beta x = \begin{cases} \beta x \, , & x \in [0, \frac{1}{\beta}) \, , \\ \beta x - 1 \, , & x \in [\frac{1}{\beta}, 1) \, . \end{cases}$$

Let $\mathcal{I} = \{ [0, \frac{1}{\beta}) , [\frac{1}{\beta}, 1) \}$ and let S be the GLS($\mathcal{I}$) transformation (see Section 2.3.1). Thus

$$Sx = \begin{cases} \beta x \, , & x \in [0, \frac{1}{\beta}) \, , \\ \beta^2 x - \beta \, , & x \in [\frac{1}{\beta}, 1) \, . \end{cases}$$

89

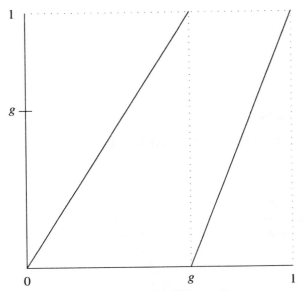

Figure 4.1. The GLS-map S

Notice that for $x \in [0, \frac{1}{\beta})$ we have $Sx = T_\beta x$ and for $x \in [\frac{1}{\beta}, 1)$,
$T_\beta x = \beta x - 1 \in [0, \frac{1}{\beta})$, while $T_\beta^2 x = T_\beta(T_\beta x) = \beta(\beta x - 1) = \beta^2 x - \beta = Sx$. Thus

$$Sx = \begin{cases} T_\beta x, & x \in [0, \frac{1}{\beta}), \\ T_\beta^2 x, & x \in [\frac{1}{\beta}, 1). \end{cases}$$

Let us now consider $\beta > 1$, where β is the positive root of the equation $f(z) = z^3 - z^2 - z - 1 = 0$. Then

$$1 = \frac{1}{\beta} + \frac{1}{\beta^2} + \frac{1}{\beta^3};$$

moreover, since $f(1) < 0$ and $f(2) > 0$ we have that $1 < \beta < 2$, which yields that $\lfloor \beta \rfloor = 1$.

Let T_β denote the β-transformation; then

$$T_\beta x = \begin{cases} \beta x, & x \in [0, \frac{1}{\beta}), \\ \beta x - 1, & x \in [\frac{1}{\beta}, 1). \end{cases}$$

Just as in Section 3.3 we let $T_\beta 1 = \beta - 1$. Consider the partition $\mathcal{I} = \{ [0, \frac{1}{\beta}), [\frac{1}{\beta}, \frac{1}{\beta} + \frac{1}{\beta^2}), [\frac{1}{\beta} + \frac{1}{\beta^2}, 1) \}$ of $[0, 1)$, and let S be the corresponding GLS($\mathcal{I}$) transformation. Thus

$$Sx = \begin{cases} \beta x & x \in [0, \frac{1}{\beta}), \\ \beta^2 x - \beta & x \in [\frac{1}{\beta}, \frac{1}{\beta} + \frac{1}{\beta^2}), \\ \beta^3 x - \beta^2 - \beta & x \in [\frac{1}{\beta} + \frac{1}{\beta^2}, 1). \end{cases}$$

Notice that

1. for $x \in [0, \frac{1}{\beta})$ we have $T_\beta x = Sx$,

2. for $x \in [\frac{1}{\beta}, \frac{1}{\beta} + \frac{1}{\beta^2})$ we have $T_\beta^2 x = \beta^2 x - \beta = Sx$, since $T_\beta x = \beta x - 1 \in [0, \frac{1}{\beta})$,

3. for $x \in [\frac{1}{\beta} + \frac{1}{\beta^2}, 1)$ we have $T_\beta^3 x = \beta(\beta^2 x - \beta - 1) = Sx$, since $T_\beta x = \beta x - 1 \in [\frac{1}{\beta}, \frac{1}{\beta} + \frac{1}{\beta^2})$ and $T_\beta^2 x = \beta^2 x - \beta - 1 \in [0, \frac{1}{\beta})$.

To summarize:

$$Sx = \begin{cases} T_\beta x, & x \in [0, \frac{1}{\beta}) = [0, T_\beta^2 1), \\ T_\beta^2 x, & x \in [\frac{1}{\beta}, \frac{1}{\beta} + \frac{1}{\beta^2}) = [T_\beta^2 1, T_\beta 1), \\ T_\beta^3 x, & x \in [\frac{1}{\beta} + \frac{1}{\beta^2}, 1) = [T_\beta 1, 1). \end{cases}$$

Here are three questions we will address in the course of this chapter:

1. Does a similar relationship exist between β-transformations and corresponding GLS transformations in general?

2. Is it possible to derive one system from the other?

3. Can we see both of these expansions as derived from one dynamical system?

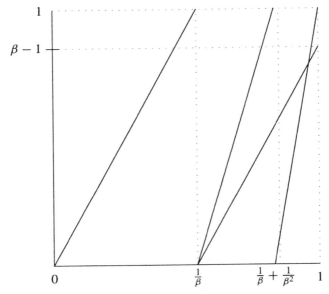

Figure 4.2. T_β and the GLS-map S

4.2 Induced and integral transformations

In 1943, S. Kakutani [Kak43] introduced the idea of a transformation
induced by a measure preserving transformation T on a subset A of
the domain of T of positive measure. The idea is to localize the system
and only observe $T^n x$ when it is in A. This has been very useful for
producing examples with a wide variety of properties. We will show
that induced transformations produce the link between β-expansions
and GLS expansions.

The basis of the construction is the Poincaré Recurrence Theorem.

Definition 4.2.1. *Let* $(X, \mathcal{F}, \mu, T)$ *be a dynamical system, (see Definition 1.2.17), and let* $B \in \mathcal{F}$. *A point* $x \in B$ *is said to be* B-*recurrent, if
there exists an integer* $k \geq 1$ *such that* $T^k x \in B$.

Theorem 4.2.2. (Poincaré Recurrence Theorem) *Let $(X, \mathcal{F}, \mu, T)$ be a dynamical system, and $B \in \mathcal{F}$ of positive measure. Then almost every $x \in B$ is B-recurrent.*

Proof. Let $F = \{x \in B : x$ is not recurrent$\}$, then

$$F = \{x \in B : T^k x \notin B, \text{ for all } k \geq 1\}.$$

Notice that $F \cap T^{-k} F = \emptyset$, for all $k \geq 1$, hence $T^{-\ell} F \cap T^{-m} F = \emptyset$, for all $\ell \neq m$. Thus the sets

$$F, \ T^{-1} F, \ T^{-2} F, \ldots$$

are mutually disjoint, and since T is measure preserving,

$$\mu(F) = \mu(T^{-1} F) = \mu(T^{-2} F) = \cdots .$$

If $\mu(F) > 0$, then

$$1 = \mu(X) \geq \mu \left(\bigcup_{k=0}^{\infty} T^{-k} F \right) = \sum_{k=0}^{\infty} \mu(T^{-k} F) = \infty,$$

which is a contradiction. ∎

In fact, the proof of Poincaré's Recurrence Theorem shows that for almost every $x \in B$, there exist infinitely many positive integers k such that $T^k x \in B$. To see this, let

$$D = \{x \in B : T^k x \in B \text{ for finitely many } k's\}$$
$$= \{x \in B : T^k x \in F \text{ for some } k \geq 0\}$$
$$= \bigcup_{k=0}^{\infty} T^{-k} F.$$

Thus $\mu(D) = 0$ since $\mu(F) = 0$ and T is measure preserving.

4.2.1   Induced transformations

Let $(X, \mathcal{F}, \mu, T)$ be a dynamical system. Let $A \subset X$ with $\mu(A) > 0$.
By Poincaré's Recurrence Theorem almost every $x \in A$ returns to A
infinitely often under the action of T. For $x \in A$, let $n(x) := \inf\{n \geq 1 : T^n x \in A\}$. We call $n(x)$ the *first return time* of x to A. Then $n(x)$
is well defined and finite a.e. on A. In the sequel we remove from A the
set of measure zero on which $n(x) = \infty$, and we denote the new set
again by A. Consider the σ-algebra $\mathcal{F} \cap A$ on A, which is the restriction
of $\mathcal{F}$ to A. Furthermore, let $\mu_A(B)$ be the probability measure on A,
defined by

$$\mu_A(B) = \frac{\mu(B)}{\mu(A)}, \quad \text{for } B \subset A,$$

so that $(A, \mathcal{F} \cap A, \mu_A)$ is a probability space. Finally, define the induced map $T_A : A \to A$ by

$$T_A x = T^{n(x)} x, \quad \text{for } x \in A.$$

From the above we see that T_A is defined on A. What kind of a transformation is T_A? We have the following proposition.

Proposition 4.2.3. T_A *is* μ_A-*invariant.*

Proof. For $k \geq 1$, let

$$A_k = \{x \in A : n(x) = k\}$$
$$B_k = \{x \in X \setminus A : Tx, \ldots, T^{k-1}x \notin A, T^k x \in A\}.$$

Notice that

$$T^{-1}A = A_1 \cup B_1 \quad \text{and} \quad T^{-1}B_n = A_{n+1} \cup B_{n+1}. \tag{4.1}$$

Let $C \in \mathcal{F} \cap A$, since T is measure preserving it follows that $\mu(C) = \mu(T^{-1}C)$.

To show that $\mu_A(C) = \mu_A(T^{-1}C)$, we show that

$$\mu(T_A^{-1}C) = \mu(T^{-1}C).$$

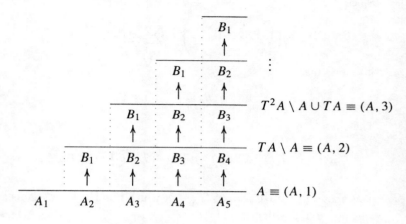

Figure 4.3. A tower

Now,

$$T_A^{-1}(C) = \bigcup_{k=1}^{\infty} A_k \cap T_A^{-1}C = \bigcup_{k=1}^{\infty} A_k \cap T^{-k}C,$$

hence

$$\mu\left(T_A^{-1}(C)\right) = \sum_{k=1}^{\infty} \mu\left(A_k \cap T^{-k}C\right).$$

On the other hand, using repeatedly (4.1), one gets for any $n \geq 1$,

$$\mu\left(T^{-1}(C)\right) = \mu(A_1 \cap T^{-1}C) + \mu(B_1 \cap T^{-1}C)$$

$$= \mu(A_1 \cap T^{-1}C) + \mu(T^{-1}(B_1 \cap T^{-1}C))$$

$$= \mu(A_1 \cap T^{-1}C) + \mu(A_2 \cap T^{-2}C) + \mu(B_2 \cap T^{-2}C)$$

$$\vdots$$

$$= \sum_{k=1}^{n} \mu(A_k \cap T^{-k}C) + \mu(B_n \cap T^{-n}C).$$

Since

$$1 \geq \mu \left(\bigcup_{n=1}^{\infty} B_n \cap T^{-n}C \right) = \sum_{n=1}^{\infty} \mu(B_n \cap T^{-n}C),$$

it follows that

$$\lim_{n \to \infty} \mu(B_n \cap T^{-n}C) = 0.$$

Thus,

$$\mu(C) = \mu(T^{-1}C) = \sum_{k=1}^{\infty} \mu \left(A_k \cap T^{-k}C \right) = \mu(T_A^{-1}C).$$

This shows that $\mu_A(C) = \mu_A(T_A^{-1}C)$, which implies that T_A is measure preserving with respect to μ_A. ∎

Exercise 4.2.4. Assume T is invertible. Show that

$$\mu_A(C) = \mu_A(T_A C),$$

for any $C \in \mathcal{F} \cap A$. (This gives another proof of Proposition 4.2.3 in the invertible case.) ∎

Exercise 4.2.5. Give a proof of the following statements:

(a) T is invertible $\Rightarrow T_A$ is invertible.
(b) T is ergodic implies that T_A is ergodic.
(c) (*Kac's Lemma*) If T is invertible and ergodic, then $\int_A n(x)d\mu = 1$.
    Conclude that $n(x) \in L^1(A, \mu_A)$, and that

$$\lim_{n \to \infty} \frac{1}{n} \sum_{i=0}^{n-1} n(T_A^i(x)) = \frac{1}{\mu(A)},$$

almost everywhere on A. ∎

Remark 4.2.6. There is a dual construction: given an induced system, we can recover the original system. We refer to this as an *integral operation*. ∎

4.2.2   Integral transformations

Given a dynamical system $(A, \mathcal{F}, \nu, S)$, let $f \in L^1(A, \nu)$ be positive and integer valued. We now construct a new dynamical system $(X, \mathcal{C}, \mu, T)$, such that the original system $(A, \mathcal{F}, \nu, S)$ is isomorphic to an induced transformation on X with return time f.

(1) $X = \{(y, i) : y \in A \text{ and } 1 \leq i \leq f(y), i \in \mathbb{N}\}$,

(2) $\mathcal{C}$ is generated by sets of the form

$$(B, i) = \{(y, i) : y \in B \text{ and } f(y) \geq i\},$$

    where $B \subset A$, $B \in \mathcal{F}$ and $i \in \mathbb{N}$.

(3) $\mu(B, i) = \dfrac{\nu(B)}{\int_A f(y) d\nu(y)}$ and then extend μ to all of X.

(4) Define $T : X \to X$ as follows:

$$T(y, i) = \begin{cases} (y, i + 1), & \text{if } i + 1 \leq f(y), \\ (Sy, 1), & \text{if } i + 1 > f(y). \end{cases}$$

Now $(X, \mathcal{C}, \mu, T)$ is called an *integral system* of $(A, \mathcal{F}, \nu, S)$ under f.

Remark 4.2.7. Suppose $(A, \mathcal{F}, \nu, S)$ is the induced dynamical system of some invertible dynamical system $(Y, \mathcal{G}, \rho, U)$ (i.e., $S = U_A$), and

$$f = n \ (= \text{first return time to } A \text{ under } U).$$

If

$$\rho\left(\bigcup_{n=0}^{\infty} U^n A\right) = 1, \tag{4.2}$$

then the integral system $(X, \mathcal{C}, \mu, T)$ of $(A, \mathcal{F}, \nu, S)$ under f is isomorphic to $(Y, \mathcal{G}, \rho, U)$. In case U is ergodic then (4.2) is satisfied.

    ■

    We now show that T is μ-measure preserving. In fact, it suffices to check this on the generators.

So let $B \subset A$ be $\mathcal{F}$-measurable, and let $i \geq 1$. We have to discern the following two cases:

(1) If $i > 1$, then $T^{-1}(B, i) = (B, i - 1)$ and clearly

$$\mu(T^{-1}(B, i)) = \mu(B, i - 1) = \mu(B, i) = \frac{v(B)}{\int_A f(y)dv(y)} \, .$$

(2) If $i = 1$, we write $A_n = \{y \in A : f(y) = n\}$, and we have

$$T^{-1}(B, 1) = \bigcup_{n=1}^{\infty} (A_n \cap S^{-1}B, \, n) \quad \text{(disjoint union)}.$$

Since $\bigcup_{n=1}^{\infty} A_n = A$ we therefore find that

$$\mu(T^{-1}(B, 1)) = \sum_{n=1}^{\infty} \frac{v(A_n \cap S^{-1}B)}{\int_A f(y)dv(y)} = \frac{v(S^{-1}B)}{\int_A f(y)dv(y)}$$

$$= \frac{v(B)}{\int_A f(y)dv(y)} = \mu(B, 1) \, .$$

This shows that T is measure preserving.

4.3   Natural extensions

We first start with the definition of a *factor*, which tells us when a dynamical system can be viewed as a subsystem of another system in the sense that the dynamical structure is preserved.

Definition 4.3.1. *Let $(X, \mathcal{F}, \mu, T)$ and $(Y, \mathcal{C}, v, S)$ be two dynamical systems. Then $(Y, \mathcal{C}, v, S)$ is said to be a factor of $(X, \mathcal{F}, \mu, T)$ if there exists a measurable and surjective map $\psi : X \to Y$ such that*

(i) *$\psi^{-1}\mathcal{C} \subset \mathcal{F}$ (so ψ preserves the measure structure);*
(ii) *$\psi T = S\psi$ (so ψ preserves the dynamics);*
(iii) *$\mu(\psi^{-1}E) = v(E) , \quad \forall E \in \mathcal{C}$ (so ψ preserves the measure).*

The dynamical system $(X, \mathcal{F}, \mu, T)$ is called an extension of $(Y, \mathcal{C}, v, S)$ and ψ is called a factor map.

Remark 4.3.2. Note that in the above definition the map ψ need not be invertible. ∎

Example 4.3.3. Let T be the Baker's transformation on $([0, 1)^2, \mathcal{B} \times \mathcal{B}, \lambda \times \lambda)$, given by

$$T(x, y) = \begin{cases} (2x, \tfrac{1}{2}y), & 0 \le x < \tfrac{1}{2} \\ (2x - 1, \tfrac{1}{2}(y + 1)), & \tfrac{1}{2} \le x < 1, \end{cases}$$

(see Exercise 1.2.22), and S the 2-ary transformation on $([0, 1), \mathcal{B}, \lambda)$, defined by

$$Sx = 2x \;(\mathrm{mod}\; 1).$$

Let $\psi : [0, 1)^2 \to [0, 1)$ be given by $\psi(x, y) = x$. It is easy to check that conditions (i), (ii), and (iii) in Definition 4.3.1 are satisfied, so that S is a factor of T. ∎

Now suppose $(Y, \mathcal{C}, v, S)$ is a *non-invertible* measure-preserving dynamical system. An invertible measure-preserving dynamical system $(X, \mathcal{F}, \mu, T)$ is called a *natural extension* of $(Y, \mathcal{C}, v, S)$ if Y is a factor of X and the factor map ψ satisfies $\bigvee_{m=0}^{\infty} T^m \psi^{-1} \mathcal{C} = \mathcal{F}$. Notice that

$$\bigvee_{k=0}^{\infty} T^k \psi^{-1} \mathcal{C}$$

is the smallest σ-algebra containing the σ-algebras $T^k \psi^{-1} \mathcal{C}$ for all $k \ge 0$.

Remarks 4.3.4.

1. There is a canonical way of constructing a natural extension of non-invertible measure preserving dynamical systems due to

V.A. Rohlin [Roh61]. Natural extensions are—up to isomorphisms—
unique. Due to this, we will refer to any natural extension of a given
transformation T as *the* natural extension $\mathcal{T}$ of T.

   2. It can be shown that the natural extension $\mathcal{T}$ is a factor of any
invertible dynamical system having T as a factor, see [Bro76], p. 26.
In this sense, $\mathcal{T}$ is the 'smallest' invertible dynamical system containing T.

   3. In many examples, the canonical construction of the natural
extension given by Rohlin may not be the easiest version to work
with. For the non-invertible transformations considered in this book,
we show how one can build a version of the natural extension in a
straightforward and convenient way. The invertible dynamical system
that we construct captures all possible "pasts" as well as "the future"
of points under the original transformation.

   4. Clearly T is ergodic if $\mathcal{T}$ is. The converse also holds, but is less
easy to prove; see e.g. [Bro76] and [CFS82].

   5. Natural extensions are extremely helpful in understanding the
dynamics of expansions. In the next section the natural extension of
GLS-expansions is given, while β-expansions are dealt with in Section
4.5. In that section we will also see that the—seemingly simple—idea
of an induced transformation is very helpful in understanding the relationship between any β-expansion and its related GLS transformation.
Similar ideas will be applied to continued fractions in Chapter 5.   ∎

Example 4.3.5. Let T^* on $\left(\{0, 1\}^{\mathbb{Z}}, \mathcal{F}, \mu\right)$ be the two-sided Bernoulli
shift, and S^* on $\left(\{0, 1\}^{\mathbb{N} \cup \{0\}}, \mathcal{G}, \nu\right)$ be the one-sided Bernoulli shift,
both Bernoulli shifts with weights $(\frac{1}{2}, \frac{1}{2})$. Notice that T^* is invertible,
while S^* is not. Set $X = \{0, 1\}^{\mathbb{Z}}$, $Y = \{0, 1\}^{\mathbb{N} \cup \{0\}}$, and define ψ :
$X \to Y$ by

$$\psi\left(\ldots, x_{-1}, x_0, x_1, \ldots\right) = (x_0, x_1, \ldots).$$

Then, ψ is a factor map. We claim that

$$\bigvee_{k=0}^{\infty} T^{*k} \psi^{-1} \mathcal{G} = \mathcal{F}.$$

To prove this, we show that $\bigvee_{k=0}^{\infty} T^{*k} \psi^{-1} \mathcal{G}$ contains all cylinders generating $\mathcal{F}$.

Let $\Delta = \{x \in X : x_{-k} = a_{-k}, \ldots, x_{\ell} = a_{\ell}\}$ be an arbitrary cylinder in $\mathcal{F}$, and let $D = \{y \in Y : y_0 = a_{-k}, \ldots, y_{k+\ell} = a_{\ell}\}$ which is a cylinder in $\mathcal{G}$. Then,

$$\psi^{-1} D = \{x \in X : x_0 = a_{-k}, \ldots, x_{k+\ell} = a_{\ell}\} \text{ and } T^{*k} \psi^{-1} D = \Delta.$$

This shows that

$$\bigvee_{k=0}^{\infty} T^{*k} \psi^{-1} \mathcal{G} = \mathcal{F}.$$

Thus, T^* is the natural extension of S^*. Notice that the maps T and S in Example 4.3.3 are isomorphic to T^* and S^* respectively (see Exercise 1.2.22). So from the above we see that T is the natural extension of S.

∎

4.4 Natural extension of the GLS transformation

Let $\mathcal{I}$ be a GLS partition, i.e., $\mathcal{I} = \{[\ell_n, r_n) : n \in \mathcal{D} \subset \mathbb{N}\}$ is a partition of $[0, 1)$ such that, if $L_n = r_n - l_n$, then $\sum_{n \in \mathcal{D}} L_n = 1$ and $0 < L_i \leq L_j < 1 \; \forall i, j \in \mathcal{D}$ with $i > j$; see also Section 2.3. Let $T : [0, 1) \to [0, 1)$ be the GLS($\mathcal{I}$) transformation. We shall define a map $\mathcal{T} : [0, 1) \times [0, 1) \to [0, 1) \times [0, 1)$ which we will prove is the natural extension of T.

Define $\mathcal{T} : [0, 1) \times [0, 1) \to [0, 1) \times [0, 1)$ as follows. Consider $x \in [0, 1)$, with GLS($\mathcal{I}$) expansion

$$x = \sum_{k=1}^{\infty} \frac{h_k}{s_1 s_2 \cdots s_k} \; ;$$

then for $y \in [0, 1)$ we define

$$\mathcal{T}(x, y) = \left(Tx, \frac{h_1}{s_1} + \frac{y}{s_1} \right).$$

Note that if $[a_1, a_2, \dots]$ is the sequence of digits associated with the GLS($\mathcal{I}$) expansion of x, and if we identify x with this sequence of digits $[a_1, a_2, \dots]$, then

$$T(x, 0) = T([a_1, a_2, \dots], 0) = ([a_2, a_3, \dots], [a_1]) ,$$

where

$$[a_2, a_3, \dots] = \frac{h_2}{s_2} + \frac{h_3}{s_2 s_3} + \cdots = \sum_{k=2}^{\infty} \frac{h_k}{s_2 \cdots s_k} \quad \text{and} \quad [a_1] = \frac{h_1}{s_1}.$$

Continuing in this way one obtains for $n \geq 1$

$$T^n(x, 0) = T^n([a_1, a_2, \dots], 0) = ([a_{n+1}, a_{n+2}, \dots], [a_n, \dots, a_1]) .$$

In general, if $y \in [0, 1)$ has GLS($\mathcal{I}$) expansion $[b_1, b_2, \dots]$, then

$$T^n(x, y) = ([a_{n+1}, a_{n+2}, \dots], [a_n, \dots, a_1, b_1, b_2, \dots]) . \qquad (4.3)$$

On $[0, 1) \times [0, 1)$ we consider two σ-algebras, the product Borel σ-algebra $\mathcal{B} \times \mathcal{B}$, and its completion $\mathcal{L} \times \mathcal{L}$. Since the inverse image under T of any rectangle is a union of rectangles, it follows that T is measurable with respect to both σ-algebras.

Proposition 4.4.1. *The transformation T is invertible and measure preserving with respect to $\lambda \times \lambda$.*

Proof. The proof of the first statement is left to the reader. For the second statement, it is enough to show that

$$(\lambda \times \lambda)(T^{-1}(\Delta_n \times \Delta_m)) = (\lambda \times \lambda)(\Delta_n \times \Delta_m)$$

for all cylinder sets Δ_n and Δ_m of rank n and m, respectively.
Let $k_1, \dots, k_n, l_1, \dots, l_m \in \mathcal{D}$; then

$$(\lambda \times \lambda)(T^{-1}(\Delta(k_1, \dots, k_n) \times \Delta(l_1, \dots, l_m)))$$

$$= (\lambda \times \lambda)(\Delta(l_1, k_1, \dots, k_n) \times \Delta(l_2, \dots, l_m))$$

$$= L_{k_1} \cdots L_{k_n} L_{l_1} \cdots L_{l_m} .$$

By Lemma 1.2.14, the proposition is true if one considers the Borel or Lebesgue σ-algebra. ∎

Theorem 4.4.2. *The system* $([0, 1) \times [0, 1), \mathcal{B} \times \mathcal{B}, \lambda \times \lambda, \mathcal{T})$ *is the natural extension of the system* $([0, 1), \mathcal{B}, \lambda, T)$, *where T and $\mathcal{T}$ are defined as above.*

Proof. Define $\pi : [0, 1) \times [0, 1) \rightarrow [0, 1)$ by $\pi(x, y) = x$. Then $\pi^{-1}[a, b) = [a, b) \times [0, 1) \in \mathcal{B} \times \mathcal{B}$ and $(\lambda \times \lambda)(\pi^{-1}[a, b)) = b - a = \lambda([a, b))$, from which we see that π is measurable and measure preserving.

It remains to show that

$$\bigvee_{m \geq 0} \mathcal{T}^m \pi^{-1} \mathcal{B} = \bigvee_{m \geq 0} \mathcal{T}^m (\mathcal{B} \times [0, 1)) = \mathcal{B} \times \mathcal{B}.$$

As in example 4.3.3, it suffices to show that $\bigvee_{m \geq 0} \mathcal{T}^m (\mathcal{B} \times [0, 1))$ contains all the two-dimensional cylinders. Referring to (4.3), we see that for any $(k_1, \ldots, k_n) \in \mathcal{D}^n$ and $(l_1, \ldots, l_m) \in \mathcal{D}^m$ one has

$$\Delta(k_1, \ldots, k_n) \times \Delta(l_1, \ldots, l_m)$$
$$= \mathcal{T}^m (\Delta(l_m, \ldots, l_1, k_1, \ldots, k_n) \times [0, 1)).$$ ∎

Again by Lemma 1.2.14, $\mathcal{T}$ is measurable and measure preserving with respect to $\mathcal{L} \times \mathcal{L}$.

Exercise 4.4.3. Define the map $\mathcal{T}_\varepsilon : [0, 1) \times [0, 1) \rightarrow [0, 1) \times [0, 1)$ by

$$\mathcal{T}_\varepsilon(x, y) := \left(T_\varepsilon x, \frac{h(x) + \varepsilon(x)}{s(x)} + \frac{(-1)^{\varepsilon(x)} y}{s(x)} \right),$$

where the map T_ε is defined as in (2.7).

(a) Show that the probability measure $\lambda \times \lambda$ is $\mathcal{T}_\varepsilon$-invariant.

(b) Show that the system $([0, 1) \times [0, 1), \mathcal{B} \times \mathcal{B}, \lambda \times \lambda, \mathcal{T}_\varepsilon)$ is the natural extension of $([0, 1), \mathcal{B}, \lambda, T_\varepsilon)$. ∎

4.5 Natural extension of the β-transformation

It is possible to give an explicit description of the natural extension of
the β-transformation for all $\beta > 1$. However, to keep things as clear as
possible, we will first describe the natural extension of T_β where β is a
pseudo-golden mean number (see also Example 3.3.4(2)). The general
case is simply a more complicated construction based on the same main
idea.

4.5.1 The pseudo-golden mean

The natural extension must be constructed so that it captures the *future*
as well as the *past* of shifts of β-expansions. It is easy—in case β is a
pseudo-golden mean number—to define a transformation on a suitable
space that does the trick. We will see that the invariant measure for the
natural extension is normalized Lebesgue measure, from which Parry's
measure as described in Section 3.4 follows. We will show that the
natural extension of the β-transformation has a suitable GLS-system
as an induced transformation. This will be used in the last result of this
book, Corollary 6.3.2.

Let $\beta > 1$ be a pseudo-golden mean number of order $m \geq 2$. That
is, $\beta > 1$ is the positive root of the polynomial $z^m - z^{m-1} - \cdots - z - 1$,
and therefore

$$1 = \frac{1}{\beta} + \frac{1}{\beta^2} + \cdots + \frac{1}{\beta^m} ,$$

so that 1 has a finite β-expansion. Denoting the β-transformation by
T_β, we then have

$$T_\beta^i 1 = \frac{1}{\beta} + \cdots + \frac{1}{\beta^{m-i}} \quad \text{for } i = 0, 1, \ldots , m - 1,$$

and $T_\beta^i 1 = 0$ for $i \geq m$. Note that in the β-expansion of any $x \in [0, 1)$,
one can have at most $m - 1$ consecutive digits equal to 1. If we view
x as a one-sided sequence of digits, then the natural extension has the
same property.

We now fix $m = 3$ just to help us visualize things; the same procedure goes through for any $m \geq 2$. As in the GLS-case—see Section 4.4—we choose as our map

$$\mathcal{T}_\beta(x, y) := \left(T_\beta x, \frac{1}{\beta}(\lfloor \beta x \rfloor + y) \right),$$

which must be defined on a proper subset X of $[0, 1)^2$. We now consider $[0, 1)$ horizontally as the future-axis, and $[0, 1)$ vertically as the past-axis, each with natural partition

$$\left[0, \frac{1}{\beta} \right), \left[\frac{1}{\beta}, \frac{1}{\beta} + \frac{1}{\beta^2} \right), \left[\frac{1}{\beta} + \frac{1}{\beta^2}, 1 \right),$$

representing points x whose β-expansion starts with a 0, a 10 or an 11 respectively. In view of this, a natural choice for X is

$$\left(\left[0, \frac{1}{\beta} \right) \times [0, 1) \right) \cup \left(\left[\frac{1}{\beta}, \frac{1}{\beta} + \frac{1}{\beta^2} \right) \times \left[0, \frac{1}{\beta} + \frac{1}{\beta^2} \right) \right)$$

$$\cup \left(\left[\frac{1}{\beta} + \frac{1}{\beta^2}, 1 \right) \times \left[0, \frac{1}{\beta} \right) \right).$$

In general one chooses

$$X = \bigcup_{k=0}^{m-1} [T_\beta^{m-k} 1, T_\beta^{m-k-1} 1) \times [0, T_\beta^k 1).$$

Notice that

$$\mathcal{T}_\beta \left(\left[0, \frac{1}{\beta} \right) \times [0, 1) \right) = [0, 1) \times \left[0, \frac{1}{\beta} \right),$$

$$\mathcal{T}_\beta \left(\left[\frac{1}{\beta}, \frac{1}{\beta} + \frac{1}{\beta^2} \right) \times \left[0, \frac{1}{\beta} + \frac{1}{\beta^2} \right) \right) = \left[0, \frac{1}{\beta} \right) \times \left[\frac{1}{\beta}, 1 \right),$$

$$\mathcal{T}_\beta \left(\left[\frac{1}{\beta} + \frac{1}{\beta^2}, 1 \right) \times \left[0, \frac{1}{\beta} \right) \right) = \left[\frac{1}{\beta}, \frac{1}{\beta} + \frac{1}{\beta^2} \right)$$

$$\times \left[\frac{1}{\beta}, \frac{1}{\beta} + \frac{1}{\beta^2} \right),$$

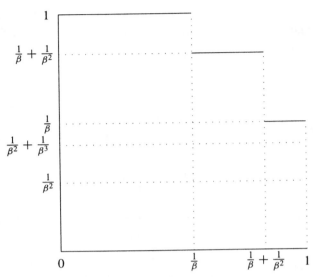

Figure 4.4.   The natural extension of T_β if $m = 3$

which shows that T_β is onto. As usual, on X we consider as σ-algebra the restriction of the product Borel or Lebesgue σ-algebras. By examining T_β, we see that both coordinates are linear maps, one with slope β, the other with slope $1/\beta$. Thus, a natural candidate for an invariant measure is the normalized Lebesgue measure on X.

Exercise 4.5.1. Let $\beta > 1$ be a pseudo-golden mean number of order $m \geq 2$, and μ normalized Lebesgue measure on X.

(a) Show that T_β is one-to-one.

(b) Show that T_β is measurable and measure preserving.   ∎

We now show that the projection of μ (which is normalized Lebesgue measure on X) on the first coordinate is the Parry measure. Let π_1 be the projection on the first coordinate, define a measure ν on $[0, 1)$ by $\nu(B) = \mu(\pi_1^{-1} B) = \mu((B \times [0, 1)) \cap X)$ for every $B \subset [0, 1)$ measurable. Then,

$$v(B) = \int_B \frac{\beta}{\frac{1}{\beta} + \frac{2}{\beta^2} + \frac{3}{\beta^3}} I_{[0,\frac{1}{\beta})}(x) \left(\frac{1}{\beta} + \frac{1}{\beta^2} + \frac{1}{\beta^3}\right) dx$$

$$+ \int_B \frac{\beta}{\frac{1}{\beta} + \frac{2}{\beta^2} + \frac{3}{\beta^3}} I_{[\frac{1}{\beta}, \frac{1}{\beta} + \frac{1}{\beta^2})}(x) \left(\frac{1}{\beta} + \frac{1}{\beta^2}\right) dx$$

$$+ \int_B \frac{\beta}{\frac{1}{\beta} + \frac{2}{\beta^2} + \frac{3}{\beta^3}} I_{[\frac{1}{\beta} + \frac{1}{\beta^2}, 1)}(x) \frac{1}{\beta} dx,$$

since

$$1 = \frac{1}{\beta} + \frac{1}{\beta^2} + \frac{1}{\beta^3}.$$

Defining $h(x)$ by

$$h(x) = \begin{cases} \dfrac{1}{\frac{1}{\beta} + \frac{2}{\beta^2} + \frac{3}{\beta^3}} \left(1 + \dfrac{1}{\beta} + \dfrac{1}{\beta^2}\right), & x \in \left[0, \dfrac{1}{\beta}\right) \\[2ex] \dfrac{1}{\frac{1}{\beta} + \frac{2}{\beta^2} + \frac{3}{\beta^3}} \left(1 + \dfrac{1}{\beta}\right), & x \in \left[\dfrac{1}{\beta}, \dfrac{1}{\beta} + \dfrac{1}{\beta^2}\right) \\[2ex] \dfrac{1}{\frac{1}{\beta} + \frac{2}{\beta^2} + \frac{3}{\beta^3}} \cdot 1, & x \in \left[\dfrac{1}{\beta} + \dfrac{1}{\beta^2}, 1\right), \end{cases}$$

it follows from above that

$$v(B) = \int_B h(x)\, dx.$$

Thus v is equivalent to Lebesgue measure λ on $[0, 1)$. Since $\pi_1 T_\beta = T_\beta \pi_1$ we see that v is a T_β-invariant measure and π_1 is a factor map. The measure v is the same measure obtained by Gelfond and Parry; see Equation (3.2) in Section 3.4. Further, by an argument similar to that used in the proof of Theorem 4.4.2, one can show that

$$\bigvee_{m=0}^{\infty} T_\beta^m \pi_1^{-1} \mathcal{B} = (\mathcal{B} \times \mathcal{B}) \cap X.$$

This shows that $(X, (\mathcal{B} \times \mathcal{B}) \cap X, \mu, \mathcal{T})$ is the natural extension of $([0, 1), \mathcal{B}, \nu, T_\beta)$. Since T_β is ergodic, see Theorem 3.4.5, it follows from Remark 4.3.4(4) that T_β is ergodic on $(X, (\mathcal{B} \times \mathcal{B}) \cap X, \mu)$, and therefore also on $(X, (\mathcal{L} \times \mathcal{L}) \cap X, \mu)$.

Let $Y = [0, 1) \times [0, 1/\beta)$ and $\mathcal{W} : Y \to Y$ be the induced transformation of T_β on Y, with return time n, given by

$$n(x, y) = \begin{cases} 1, & (x, y) \in [0, \frac{1}{\beta}) \times [0, \frac{1}{\beta}), \\ 2, & (x, y) \in [\frac{1}{\beta}, \frac{1}{\beta} + \frac{1}{\beta^2}) \times [0, \frac{1}{\beta}), \\ 3, & (x, y) \in [\frac{1}{\beta} + \frac{1}{\beta^2}, 1) \times [0, \frac{1}{\beta}). \end{cases}$$

One easily sees that $\mathcal{W}$ is given by

$$\mathcal{W}(x, y) = \begin{cases} (\beta x, \frac{y}{\beta}), & (x, y) \in [0, \frac{1}{\beta}) \times [0, \frac{1}{\beta}), \\ (\beta^2 x - \beta, \frac{1}{\beta^2}(1 + y)), & (x, y) \in [\frac{1}{\beta}, \frac{1}{\beta} + \frac{1}{\beta^2}) \times [0, \frac{1}{\beta}), \\ (\beta^3 x - \beta^2 - \beta, \frac{1}{\beta^2} + \frac{1}{\beta^3} + \frac{y}{\beta^4}), & \\ & (x, y) \in [\frac{1}{\beta} + \frac{1}{\beta^2}, 1) \times [0, \frac{1}{\beta}), \end{cases}$$

and that the invariant measure is $\beta(\lambda \times \lambda)$.

We now show that the induced system $(Y, \mathcal{W})$ is isomorphic to the natural extension of a GLS($\mathcal{I}$) operator in case we use the Borel σ-algebra on both spaces, and is isomorphic to the completion of the GLS($\mathcal{I}$) operator in case we use the Lebesgue σ-algebra. Let $\mathcal{I}$ be given by

$$\mathcal{I} = \left\{ \left[0, \frac{1}{\beta}\right), \left[\frac{1}{\beta}, \frac{1}{\beta} + \frac{1}{\beta^2}\right), \left[\frac{1}{\beta} + \frac{1}{\beta^2}, 1\right) \right\}.$$

Then $\mathcal{I}$ is a GLS partition. Let S be the corresponding GLS operator. Thus,

$$Sx = \begin{cases} \beta x, & 0 \le x < \frac{1}{\beta}, \\ \beta^2 x - \beta, & \frac{1}{\beta} \le x < \frac{1}{\beta} + \frac{1}{\beta^2}, \\ \beta^3 x - \beta^2 - \beta, & \frac{1}{\beta} + \frac{1}{\beta^2} \le x < 1. \end{cases}$$

Let $\mathcal{S}$ be the natural extension map of S, so that

$$
S(x, y) = \begin{cases}
\left(\beta x, \dfrac{y}{\beta}\right), & (x, y) \in \left[0, \dfrac{1}{\beta}\right) \times [0, 1), \\[2ex]
\left(\beta^2 x - \beta, \dfrac{1}{\beta} + \dfrac{y}{\beta^2}\right), & (x, y) \in \left[\dfrac{1}{\beta}, \dfrac{1}{\beta} + \dfrac{1}{\beta^2}\right) \times [0, 1), \\[2ex]
\left(\beta^3 x - \beta^2 - \beta, \dfrac{1}{\beta} + \dfrac{1}{\beta^2} + \dfrac{y}{\beta^3}\right), & \\[2ex]
& (x, y) \in \left[\dfrac{1}{\beta} + \dfrac{1}{\beta^2}, 1\right) \times [0, 1).
\end{cases}
$$

We have seen that $\mathcal{S}$ is measure preserving with respect to $\lambda \times \lambda$ on $[0, 1)^2$. The following proposition is easily verified.

Proposition 4.5.2. *Let* $\phi : [0, 1)^2 \to Y$ *be defined by* $\phi(x, y) = (x, \frac{y}{\beta})$. *Then* ϕ *is an isomorphism from* $\left([0, 1)^2, \lambda \times \lambda, \mathcal{S}\right)$ *to* $(Y, \beta(\lambda \times \lambda), \mathcal{W})$.

4.5.2   General β

The general case is a more complicated version of the pseudo-golden mean case. Our aim is to build an invertible dynamical system that captures the past as well as the future of the map T_β and is the smallest dynamical system with such a property, in the sense that it is a factor of any other dynamical system that also captures the past and future of T_β. In case β belongs to a special class of algebraic numbers, other versions of the natural extension can be also obtained; see [Nak95].

We will outline briefly the construction of the natural extension, and the related GLS as an induced system; see also [DKS96].

Let

$$
R_0 = [0, 1)^2 \text{ and } R_i = [0, T_\beta^i 1) \times \left[0, \dfrac{1}{\beta^i}\right), \ i \geq 1;
$$

the underlying space $\mathcal{H}_\beta$ is obtained by stacking (as pages in a book) R_{i+1} on top of R_i, for each $i \geq 0$. The index i indicates at what

height one is in the stack. (In case 1 has a finite β-expansion of length
n, only n R_i's are stacked.) Let $\mathcal{B}_i$ be the collection of Borel sets of
R_i, and let the σ-algebra $\mathcal{F}$ on $\mathcal{H}_\beta$ be the direct sum of the $\mathcal{B}_i$'s, i.e.,
$\mathcal{F} = \oplus \mathcal{B}_i$. Furthermore, the measure on $\mathcal{H}_\beta$ that is Lebesgue measure
on each rectangle R_i is denoted by η, and we put $\mu = \frac{1}{\eta(\mathcal{H}_\beta)}\eta$. Finally
$T_\beta : \mathcal{H}_\beta \to \mathcal{H}_\beta$ is defined as follows. Let $d^*(\beta) = .b_1 b_2 \ldots$ (as in
Proposition 3.3.3), and let $(x, y) \in R_i$, $i \geq 0$, where $x = .d_1 d_2 \ldots$ is
the β-expansion of x and $y = . \underbrace{00 \ldots 0}_{i-\text{times}} c_{i+1} c_{i+2} \cdots$ is the β-expansion
of y (notice that $(x, y) \in R_i$ implies that $d_1 \leq b_{i+1}$). In view of Proposition 3.3.3, we now define

$$T_\beta(x, y) := (T_\beta x, y^*) \in \begin{cases} R_0, & \text{if } d_1 < b_{i+1}, \\ R_{i+1}, & \text{if } d_1 = b_{i+1}, \end{cases} \qquad (4.4)$$

where

$$y^* = \begin{cases} \dfrac{b_1}{\beta} + \cdots + \dfrac{b_i}{\beta^i} + \dfrac{d_1}{\beta^{i+1}} + \dfrac{y}{\beta} = .b_1 \cdots b_i d_1 c_{i+1} c_{i+2} \cdots, \\ \qquad\qquad\qquad\qquad\qquad\qquad\qquad \text{if } d_1 < b_{i+1}, \\ \dfrac{y}{\beta} = .\underbrace{000 \ldots 00}_{i+1-\text{times}} c_{i+1} c_{i+2} \cdots, \qquad\qquad \text{if } d_1 = b_{i+1}. \end{cases}$$

Notice that in case $i = 0$ one has

$$y^* = \begin{cases} \dfrac{1}{\beta}(y + d_1), & d_1 < b_1, \\ \dfrac{y}{\beta}, & d_1 = b_1. \end{cases}$$

Exercise 4.5.3. Show that the Parry measure as given in Equation (3.2)
is obtained by projecting μ on the first coordinate.   ∎

Consider the probability space $\left(R_0, \mathcal{H}_\beta \cap R_0, \mu_{R_0}\right)$, which by
construction equals $\left([0, 1)^2, \mathcal{B} \times \mathcal{B}, \lambda \times \lambda\right)$. Let $\mathcal{W}_\beta : R_0 \to R_0$
be the induced map of T_β on R_0. That is, $\mathcal{W}_\beta$ is given as follows.

Let $(x, y) \in R_0$, where $x = .d_1 d_2 \ldots$ is the β-expansion of x and $d^*(\beta) = .b_1 b_2 \ldots$; then

$$W_\beta(x, y) = \begin{cases} T_\beta(x, y), & \text{if } d_1 < b_1, \\ T_\beta^n(x, y), & \text{if } d_i = b_i \text{ for } 1 \le i \le n-1 \text{ and } d_n < b_n. \end{cases}$$

This follows from definition (4.4) of T_β on R_0, for

(i) if $d_1 < b_1$ then $T_\beta(x, y) \in R_0$.

(ii) if $d_i = b_i$ for $1 \le i \le n-1$ and $d_n < b_n$ one has $T_\beta^i(x, y) \in R_i$ for $1 \le i \le n-1$ and $T_\beta^n(x, y) \in R_0$.

For $(x, y) \in R_0$, let $n(x, y) := \inf\{k \ge 1 ; T_\beta^k(x, y) \in R_0\}$ and $R_0^k := \{(x, y) \in R_0 ; n(x, y) = k\}$; thus

$$W_\beta(x, y) = \begin{cases} \left(T_\beta x, \dfrac{1}{\beta}(y + d_1)\right), & (x, y) \in R_0^1, \\[3mm] \left(T_\beta^k x, \dfrac{b_1}{\beta} + \cdots + \dfrac{b_{k-1}}{\beta^{k-1}} + \dfrac{d_k}{\beta^k} + \dfrac{y}{\beta^k}\right), & \\ & (x, y) \in R_0^k, \ k \ge 2. \end{cases}$$

Notice that $n(x, y)$ is the first return time of the map T_β to R_0, where $n \in \mathbb{N}$. The above shows that

$$R_0^1 = [0, .b_1) \times [0, 1)$$
$$R_0^k = [.b_1 \cdots b_{k-1}, .b_1 \cdots b_{k-1} b_k) \times [0, 1), \ k \ge 2, \qquad \text{and}$$
$$R_0 = \bigcup_{k=1}^\infty R_0^k.$$

Let $\mathcal{I} = \{I_n ; n \ge 1\}$ be the partition on $[0, 1)$ defined as follows: for each $n \ge 1$, there exist unique integers $k \ge 0$ and $1 \le i \le b_{k+1}$ such that

$$n = b_0 + b_1 + \cdots + b_k + (i - 1),$$

where by definition we put $b_0 := 0$. Set

$$I_n = \left[b_0 + \frac{b_1}{\beta} + \cdots + \frac{b_k}{\beta^k} + \frac{i-1}{\beta^{k+1}}, \; b_0 + \frac{b_1}{\beta} + \cdots + \frac{b_k}{\beta^k} + \frac{i}{\beta^{k+1}} \right).$$

$$(4.5)$$

Note that for $(x, y) \in I_n \times [0, 1)$,

$$\mathcal{W}_\beta(x, y) = T_\beta^{k+1}(x, y)$$

$$= \left[T_\beta^{k+1} x, \; b_0 + \frac{b_1}{\beta} + \cdots + \frac{b_k}{\beta^k} + \frac{i-1}{\beta^{k+1}} + \frac{y}{\beta^{k+1}} \right).$$

We have the following proposition.

Proposition 4.5.4. *Let $\mathcal{I} = (I_n)_{n\in\mathbb{N}}$ be the partition from (4.5). Then the natural extension $\mathcal{T}$ of the $\mathrm{GLS}(\mathcal{I})$ transformation T is identical to $\mathcal{W}_\beta$.*

Proof. The proof follows from the fact that $\mathcal{T}$ on $I_{b_0 + \cdots + b_k + (i-1)} \times [0, 1)$ for $1 \le i \le b_{k+1}$ is given by

$$\mathcal{T}(x, y) = \left(Tx, \; b_0 + \frac{b_1}{\beta} + \cdots + \frac{b_k}{\beta^k} + \frac{i-1}{\beta^{k+1}} + \frac{y}{\beta^{k+1}} \right).$$

Since for $x \in [b_0 + \frac{b_1}{\beta} + \cdots + \frac{b_k}{\beta^k}, \; b_0 + \frac{b_1}{\beta} + \cdots + \frac{b_{k+1}}{\beta^{k+1}})$, $Tx = T_\beta^{k+1}x$, it follows that $\mathcal{W}_\beta(x, y) = \mathcal{T}(x, y)$.

Remark 4.5.5. By Remarks 4.3.4(2) we see that $\mathcal{T}$, and hence $\mathcal{W}_\beta$, are ergodic. ∎

4.6   For further reading

Several good books on ergodic theory have been mentioned in this chapter, or previously. Our first choices of reference and further reading are *An Introduction to Ergodic Theory* by Peter Walters [Wal82], and *Ergodic Theory* by Karl Petersen [Pet89]. We also recommend Billingsley's book [Bil65], although it is old now and most of the open problems mentioned in it—like the isomorphism problem for Bernoulli Shifts—have been solved.

A book of encyclopedic nature by I.P. Cornfeld, S.V. Fomin and Ya.G. Sinai [CFS82] is unfortunately handicapped by a largely non-existing index; moreover it is quite hard for a beginner.

Other excellent books are (in a random order) by U. Krengel [Kre85], J.R. Brown [Bro76], M. Pollicott and M. Yuri [PY98], and G. Keller [Kel98]. Finally the books by R. Mañé [Mn87] and D.J. Rudolph [Rud90] should not be omitted from this list, but are not first recommendations for beginners.

For topics not mentioned here, we recommend H. Furstenberg's beautiful monograph [Fur81], where an introduction to recurrence in topological dynamical systems is given. What makes Furstenberg's book particularly interesting is that a multidimensional version of Szemeredi's theorem is obtained on the existence of arbitrarily long arithmetic progressions in sequences of integers with positive density.

CHAPTER 5

Diophantine approximation and continued fractions

5.1 Introduction

5.1.1 Why continued fractions?

In Section 1.3.1 continued fractions were introduced as a generalization of Euclid's algorithm. Although at first view they appear strange, continued fractions play an important role at many places in mathematics. For instance they appear in *primality testing*, which is not so surprising, since Euclid's algorithm is also a test whether two integers m and n are relatively prime. For more details, see the book by Bressoud [Bre89]. There is also an important relation between continued fractions and *algebraic geometry*, see the nice introductory papers by C. Series [Ser82] and [Ser85].

Another important application of continued fractions is the *approximation* of real irrational numbers by rationals, also known as *Diophantine approximation*, which is a phrase derived from Diophantus of Alexandria, who lived around AD 250. In this chapter we will go deeper into this, and we will find many approximation results as spin-offs of the natural extension of the regular continued fraction. Before doing so, we will first recall some classical results by Dirichlet and Hurwitz.

Theorem 5.1.1. (Dirichlet, 1842) *Let x and Q be real numbers, where $Q > 1$. Then there exist integers p and q, with $1 \leq q < Q$ such that*

$$\left| x - \frac{p}{q} \right| \leq \frac{1}{qQ} . \tag{5.1}$$

Proof. Notice that we may assume that Q is an integer, for otherwise we may replace Q by Q^*, defined by $Q^* := \lfloor Q \rfloor + 1$.

Essential in the proof of Dirichlet's theorem is the so-called *pigeon-hole principle* ('if you want to put $n + 1$ letters in n boxes, at least one box will contain at least two letters'); the following $Q + 1$ numbers all lie in the unit interval $[0, 1]$:

$$0, \quad \{x\}, \quad \{2x\}, \dots, \quad \{(Q-1)x\}, \quad 1 ; \tag{5.2}$$

here $\{\xi\}$ denotes the *fractional part* of ξ, i.e., $\{\xi\} := \xi - \lfloor \xi \rfloor$. Now partition $[0, 1]$ into Q subintervals of equal length $1/Q$:

$$\left[0, \frac{1}{Q} \right), \quad \left[\frac{1}{Q}, \frac{2}{Q} \right), \dots, \quad \left[\frac{Q-2}{Q}, \frac{Q-1}{Q} \right), \quad \left[\frac{Q-1}{Q}, 1 \right] . \tag{5.3}$$

Then there is at least one of these subintervals that contains two (or more) of the $Q + 1$ numbers from (5.2), i.e., there exist two integers q_1 and q_2 with $q_1 \neq q_2$ (say $q_1 > q_2$) and $1 \leq q_1, q_2 < Q$, such that both $\{q_1 x\}$ and $\{q_2 x\}$ are contained in the same subinterval from (5.3). Since

$$\{q_i x\} = q_i x - p_i , \quad i = 1, 2,$$

for some appropriate $p_1, p_2 \in \mathbb{Z}$, it follows that

$$|(q_1 x - p_1) - (q_2 x - p_2)| \leq \frac{1}{Q} .$$

Setting

$$q := q_1 - q_2, \quad p := p_1 - p_2,$$

we have $p, q \in \mathbb{Z}$, $1 \leq q < Q$ and

$$|qx - p| \leq \frac{1}{Q},$$

which proves the theorem. ∎

Exercise 5.1.2. Show that in case x is a real irrational number, there exist infinitely many pairs of integers p and q, with $q \geq 1$ and $(p, q) = 1$, such that

$$\left| x - \frac{p}{q} \right| \leq \frac{1}{q^2}. \tag{5.4}$$

∎

Exercise 5.1.3. Show that the result in Exercise 5.1.2 is never true in case x is rational. ∎

In 1891, Hurwitz showed that for irrational numbers x inequality (5.4) can be improved considerably.

Theorem 5.1.4. (Hurwitz, 1891) *For every irrational number x there exist infinitely many pairs of integers p and q, such that*

$$\left| x - \frac{p}{q} \right| \leq \frac{1}{\sqrt{5}} \frac{1}{q^2}. \tag{5.5}$$

The constant $1/\sqrt{5}$ is best possible; i.e., if we replace $1/\sqrt{5}$ by a smaller constant C, then there are infinitely many irrational numbers x for which

$$\left| x - \frac{p}{q} \right| \leq \frac{C}{q^2}$$

holds for only finitely many pairs of integers p and q.

There are several ways to prove this theorem. One classical way is to use Ford circles; see also the books by W.M. Schmidt [Sch80b] and Rademacher [Rad83]. Another way is to use continued fractions. In 1903, Borel obtained the following theorem, which at once implies Hurwitz' theorem.

Theorem 5.1.5. (Borel, 1903) *Let $n \geq 1$ and let $\frac{p_{n-1}}{q_{n-1}}$, $\frac{p_n}{q_n}$ and $\frac{p_{n+1}}{q_{n+1}}$ be three consecutive continued fraction convergents to the irrational number x. Then at least one of these three convergents satisfies* (5.5).

In the 80 years following Borel's result several generalizations and improvements to this theorem have been obtained. In Section 5.3.1 we will show that a dynamical approach to continued fractions yields these generalizations and improvements as easy corollaries.

5.1.2 Approximation coefficients

In Exercise 5.1.2 we saw that for x irrational

$$\left| x - \frac{p}{q} \right| \leq \frac{1}{q^2}$$

has infinitely many rational solutions $\frac{p}{q}$. From Borel's Theorem 5.1.5 it follows that infinitely many of these solutions are (regular) continued fraction convergents $\frac{p_n}{q_n}$ of x. A classical theorem of Legendre shows that the real good rational approximations $\frac{p}{q}$ of x are in fact always convergents $\frac{p_n}{q_n}$. We will present here a refined version of Legendre's Theorem by D. Barbolosi and H. Jager [BJ94]. We first give some definitions.

Definition 5.1.6. *The approximation coefficient $\Theta(x, \frac{p}{q})$ of a rational number $\frac{p}{q}$ with respect to a real irrational number x is defined by*

$$\Theta\left(x, \frac{p}{q}\right) = q|qx - p|,$$

i.e.,

$$\left| x - \frac{p}{q} \right| = \frac{\Theta(x, p/q)}{q^2}.$$

In case $\frac{p}{q}$ equals the kth continued fraction convergent $\frac{p_k}{q_k}$ of x for some k, we write $\frac{p}{q} \in \mathrm{RCF}(x)$, and we write $\Theta_k(x)$ (or Θ_k) instead of $\Theta(x, \frac{p}{q})$. Further we define

$$\epsilon\left(x, \frac{p}{q}\right) = \begin{cases} +1, & \text{if } x > \frac{p}{q}, \\ -1, & \text{if } x < \frac{p}{q}. \end{cases}$$

Finally, the signature $\delta(x, \frac{p}{q})$ of $\frac{p}{q}$ with respect to x is defined by

$$\delta\left(x, \frac{p}{q}\right) = (-1)^n \epsilon\left(x, \frac{p}{q}\right),$$

where n is the depth of the continued fraction expansion of $\frac{p}{q}$. (See also Section 1.3.1.)

Note that Hurwitz' Theorem 5.1.4, Borel's Theorem 5.1.5 and Exercise 5.1.2 are in fact statements on approximation coefficients, since in general

$$\left| x - \frac{p}{q} \right| = \frac{\Theta(x, p/q)}{q^2},$$

and in particular

$$\left| x - \frac{p_k}{q_k} \right| = \frac{\Theta_k}{q_k^2}.$$

Given an irrational x and a rational p/q, how can we determine in a quick and efficient manner whether $p/q \in \mathrm{RCF}(x)$, without first expanding x in a continued fraction? In the second year of the French Revolution, Legendre gave a partial answer in terms of the approximation coefficients, see Corollary 5.1.8. In 1994, Barbolosi and Jager gave a complete characterization by generalizing Legendre's result.

Theorem 5.1.7. (Barbolosi and Jager, 1994) *Let p and q be two integers such that $(p, q) = 1$, $q > 0$, and let x be a real irrational number.*

If $\delta(x, \frac{p}{q}) = +1$, then

$$\Theta\left(x, \frac{p}{q}\right) < \frac{2}{3} \;\Rightarrow\; \frac{p}{q} \in \mathrm{RCF}(x),$$

and

$$\Theta\left(x, \frac{p}{q}\right) > 1 \;\Rightarrow\; \frac{p}{q} \notin \mathrm{RCF}(x).$$

If on the other hand $\delta(x, \frac{p}{q}) = -1$, then

$$\Theta\left(x, \frac{p}{q}\right) < \frac{1}{2} \;\Rightarrow\; \frac{p}{q} \in \mathrm{RCF}(x),$$

and

$$\Theta\left(x, \frac{p}{q}\right) > \frac{2}{3} \;\Rightarrow\; \frac{p}{q} \notin \mathrm{RCF}(x).$$

All constants are best possible, i.e., cannot be replaced by bigger constants.

Proof. Suppose that the *depth n* of $\frac{p}{q}$ is even (the case n odd runs along lines similar to the even case, and is therefore omitted). That is, $\frac{p}{q} = [a_0; a_1, \ldots, a_n]$ with $a_n \geq 2$, and n is even. Denote by $\frac{r}{s}$ the last but one convergent of $\frac{p}{q}$, i.e., $\frac{r}{s} = [a_0; a_1, \ldots, a_{n-1}]$.

The set of all irrational numbers x with $x > \frac{p}{q}$, i.e., with $\delta(x, \frac{p}{q}) = +1$, and with $\frac{p}{q} \in \mathrm{RCF}(x)$ is just the cylinder set $\Delta^* = \Delta_n(a_1, \ldots, a_n) \setminus \mathbb{Q}$ of order n. For any irrational x in this cylinder Δ^* one has that

$$\frac{p}{q} = \frac{p_n}{q_n} \quad \text{and} \quad \frac{r}{s} = \frac{p_{n-1}}{q_{n-1}},$$

and from Exercise 1.3.15 it follows that Δ^* equals

$$\left(\frac{p}{q}, \frac{p+r}{q+s}\right) \setminus \mathbb{Q},$$

which has length $q^{-1}(q+s)^{-1}$, since $rq - sp = p_{n-1}q_n - p_n q_{n-1} = 1$;
see Exercise 1.3.8. From $\delta(x, \frac{p}{q}) = +1$ we have

$$x \in \Delta^* \quad \Leftrightarrow \quad \Theta\left(x, \frac{p}{q}\right) = q^2 \left| x - \frac{p}{q} \right| < \frac{q}{q+s}.$$

Thus

$$\frac{p}{q} \in \mathrm{RCF}(x) \quad \Leftrightarrow \quad \Theta\left(x, \frac{p}{q}\right) < \frac{1}{1 + \frac{s}{q}}. \tag{5.6}$$

By Exercise 1.3.9, $\frac{s}{q} = [0; a_n, \dots, a_1] \in \Delta_1(a_n)$ with $\Delta_1(a_n)$ is an
interval with endpoints $\frac{1}{a_n+1}$ and $\frac{1}{a_n}$; thus we have

$$\frac{1}{a_n + 1} \le \frac{s}{q} < \frac{1}{a_n}. \tag{5.7}$$

Therefore,

$$\Theta\left(x, \frac{p}{q}\right) < \frac{a_n}{a_n + 1} \quad \Rightarrow \quad \frac{p}{q} \in \mathrm{RCF}(x)$$

and

$$\Theta\left(x, \frac{p}{q}\right) > \frac{a_n + 1}{a_n + 2} \quad \Rightarrow \quad \frac{p}{q} \notin \mathrm{RCF}(x).$$

Since

$$\frac{a_n}{a_n + 1} \ge \frac{2}{3} \quad \text{and} \quad \frac{a_n + 1}{a_n + 2} \le 1,$$

for all $a_n \ge 2$, the constants cannot be replaced by bigger ones, and the
first part of the theorem follows.

Let $x < \frac{p}{q}$, and notice that $\frac{p}{q} = [a_0; a_1, \dots, a_n - 1, 1]$. By
Exercise 1.3.15 the set of all irrational numbers x with $x < \frac{p}{q}$ and
$\frac{p}{q} \in \mathrm{RCF}(x)$ is the cylinder set $\Delta^\sharp = \Delta_{n+1}(a_1, \dots, a_n - 1, 1) \setminus \mathbb{Q}$ of
order $n + 1$, with endpoints $\frac{p_n^\sharp}{q_n^\sharp}$ and $\frac{p_{n+1}^\sharp}{q_{n+1}^\sharp}$, with

$$\frac{p_n^\sharp}{q_n^\sharp} = [a_0; a_1, \dots, a_n - 1]$$

and

$$\frac{p_{n+1}^{\sharp}}{q_{n+1}^{\sharp}} = [a_0; a_1, \ldots, a_n - 1, 1] = \frac{p}{q}.$$

From the recursion relations (1.12) it follows that

$$p_n^{\sharp} = p_n - p_{n-1} = p - r,$$
$$q_n^{\sharp} = q_n - q_{n-1} = q - s,$$

so from Exercise 1.3.15 it follows that

$$\Delta^{\sharp} = \left(\frac{2p-r}{2q-s}, \frac{p}{q}\right) \setminus \mathbb{Q},$$

which has length $q^{-1}(2q-s)^{-1}$. A similar argument leading to (5.6) now yields

$$\frac{p}{q} \in \mathrm{RCF}(x) \Leftrightarrow \Theta(x, \frac{p}{q}) < \frac{1}{2 - \frac{s}{q}}. \tag{5.8}$$

Notice that (5.7) is equivalent to

$$\frac{a_n + 1}{2a_n + 1} \leq \frac{1}{2 - \frac{s}{q}} < \frac{a_n}{2a_n - 1},$$

and therefore the assertions follow again, since

$$\frac{a_n + 1}{2a_n + 1} \geq \frac{1}{2} \quad \text{and} \quad \frac{a_n}{2a_n - 1} \leq \frac{2}{3}$$

for all $a_n \geq 2$. As above we see the constants are best possible. This proves the theorem. ∎

Legendre's theorem now follows from the result of Barbolosi and Jager as a corollary.

Corollary 5.1.8. (Legendre) *Let p and q be two integers such that $(p, q) = 1$, $q > 0$, and let x be a real irrational number. Then*

$$\Theta\left(x, \frac{p}{q}\right) < \frac{1}{2} \Rightarrow \frac{p}{q} \in \mathrm{RCF}(x).$$

The constant $\frac{1}{2}$ is best possible.

Remark 5.1.9. Let $x \in [0, 1)$ be an irrational number with (regular) continued fraction expansion $x = [0; a_1, \ldots, a_n, \ldots]$, with sequence of convergents $(\frac{p_n}{q_n})_{n \geq 0}$, and with approximation coefficients $\Theta_n = \Theta_n(x)$, $n \geq 0$. Barbolosi and Jager also characterized the rationals $\frac{p}{q}$ that are *not* convergents but still satisfy (5.4). They showed that these $\frac{p}{q}$ are from the set of *intermediate convergents* of x, defined by

$$\frac{kp_n + p_{n-1}}{kq_n + q_{n-1}}, \quad 1 \leq k \leq a_{n+1} - 1, \, n \geq 0.$$

For more details, see [BJ94]. ∎

Remark 5.1.10. In Section 1.3.1 we abbreviated $T^n(x)$ by T_n for $n \geq 1$. Notice that

$$T_n = [0; a_{n+1}, a_{n+2}, \ldots],$$

which can be understood as the future of x at time n. Similarly we define

$$V_n = \frac{q_{n-1}}{q_n} \quad \text{for } n \geq 1 \text{ and } V_0 = 0.$$

Thus

$$V_n = [0; a_n, \ldots, a_1], \, n \geq 1; \, V_0 = 0,$$

is the past of x at time n. ∎

Exercise 5.1.11. Show, using (1.12), that for $n \geq 1$ one has

$$\Theta_{n-1} = \frac{V_n}{1 + T_n V_n} \quad \text{and} \quad \Theta_n = \frac{T_n}{1 + T_n V_n}. \tag{5.9}$$

∎

From (5.9) it at once follows that each approximation coefficient Θ_n is a number between 0 and 1. Roughly speaking, this means that the speed with which p_n/q_n converges to x is the same as the speed with which the denominators q_n (and hence also the numerators p_n) tend to infinity. The recurrence relations (1.12) show that this convergence is at least as fast as the way the Fibonacci numbers grow (these are the q_n's one gets if all a_n's are equal to 1). However, due to Lévy's Proposition, Proposition 3.5.5, we see that this convergence is exponential for a.e. x.

With a little effort one finds much more information on the approximation coefficients Θ_n. For each irrational x one clearly has that

$$(T_n, V_n) \in [0, 1) \times [0, 1], \quad \text{for } n \geq 0.$$

With $\Omega := [0, 1) \times [0, 1]$, in view of (5.9) it seems natural to study the map $\psi : \Omega \to \mathbb{R}^2$, defined by

$$\psi(x, y) := \left(\frac{y}{1 + xy}, \frac{x}{1 + xy} \right), \quad \text{for } (x, y) \in \Omega. \tag{5.10}$$

In this case one has for all irrational numbers x and all $n \geq 1$ that

$$(\Theta_{n-1}, \Theta_n) = \psi(T_n, V_n) \in \psi(\Omega). \tag{5.11}$$

Exercise 5.1.12. Let $\Gamma := \psi(\Omega)$. Show that Γ is the triangle in $\mathbb{R}^2$ with vertices $(0, 0)$, $(0, 1)$ and $(1, 0)$, specify its boundary and show that $\psi : \Omega \to \Gamma$ is a bijection.   ∎

From (5.11) and Exercise 5.1.12 the following result by Vahlen follows at once.

Corollary 5.1.13. (Vahlen, 1895) *For all irrational numbers x and all $n \geq 1$ one has*

$$\min(\Theta_{n-1}, \Theta_n) < \frac{1}{2}.$$

In order to obtain Borel's Theorem, Theorem 5.1.5, and its generalizations and refinements, one needs to have a better understanding of the dynamics of the sequence $((\Theta_{n-1}, \Theta_n))_{n\geq 1}$ in $\Gamma = \psi(\Omega)$ for (almost) all x. In order to do so, we study Ω in the next section as the space for the *natural extension* of the (regular) continued fraction. In Section 5.3 we will finally derive the generalizations of Borel's Theorem, by going from Ω to Γ.

5.2   The natural extension of the regular continued fraction

In Section 1.3.3 we saw that the probability measure μ on $[0, 1)$, with density function

$$\frac{1}{\log 2} \frac{1}{1+x}, \quad x \in [0, 1),$$

the so-called *Gauss measure*, is the invariant measure for the continued fraction map T. In Section 3.5, Theorem 3.5.1, it was shown that $([0, 1), \mathcal{B}, \mu, T)$ is an ergodic system.

In order to study the distribution of the sequence $(\Theta_n(x))_{n\geq 1}$ for almost all x (with respect to Lebesgue measure), the system $([0, 1), \mu, T)$ is 'insufficient', since it only deals with the future, while Θ_n depends on both past V_n and future T_n, as we saw in (5.9). Since $0 \leq T_n < 1$ for $x \notin \mathbb{Q}$ and $0 \leq V_n \leq 1$, we introduced in the previous section the space $\Omega = [0, 1) \times [0, 1]$. Now define a map $\mathcal{T} : \Omega \to \Omega$ by

$$\mathcal{T}(x, y) := \left(Tx, \frac{1}{a_1(x) + y} \right), \quad (x, y) \in \Omega.$$

Exercise 5.2.1. Show that for each irrational number $x \in [0, 1)$ and for all $n \geq 0$ one has

$$\mathcal{T}^n(x, 0) = (T_n, V_n).$$ ∎

We have the following theorem.

Theorem 5.2.2. (Nakada, Ito and Tanaka, [NIT77], [Nak81]) *Let $\bar{\mu}$ be the probability measure on Ω with density $d(x, y)$, given by*

$$d(x, y) := \frac{1}{\log 2} \frac{1}{(1 + xy)^2}, \quad (x, y) \in \Omega;$$

then $\bar{\mu}$ is the invariant measure for $\mathcal{T}$. Furthermore, the dynamical system

$$(\Omega, \bar{\mu}, \mathcal{T})$$

is an ergodic system.

Exercise 5.2.3. Use ideas from Section 4.4 to prove Theorem 5.2.2. ∎

5.3 Approximation coefficients revisited

5.3.1 Arithmetical properties

In this section we will prove—via several exercises—the following theorem.

Theorem 5.3.1. *Let x be an irrational number, and let $n \geq 1$. Then*

$$\min(\Theta_{n-1}, \Theta_n, \Theta_{n+1}) < \frac{1}{\sqrt{a_{n+1}^2 + 4}} \tag{5.12}$$

and

$$\max(\Theta_{n-1}, \Theta_n, \Theta_{n+1}) > \frac{1}{\sqrt{a_{n+1}^2 + 4}}. \tag{5.13}$$

Remark 5.3.2. Clearly (5.12) generalizes Borel's Theorem (Theorem 5.1.5). Several authors have given various proofs of (5.12), e.g., Bagemihl and McLaughlin [BM66]. Inequality (5.13) was first obtained by J. Tong [Ton83] in 1983. The approach in this section is different from all other proofs, and was first published in [JK89]. It can also be used to obtain asymmetric results, as first found by Segre [Seg45] in 1945. For such results the interested reader is referred to [Kra90].   ∎

Exercise 5.3.3. Using the notation from (5.10) and Exercise 5.1.12, show that the inverse $\psi^{-1} : \Gamma \to \Omega$ is given by

$$\psi^{-1}(\alpha, \beta) = \left(\frac{1 - \sqrt{1 - 4\alpha\beta}}{2\alpha}, \frac{1 - \sqrt{1 - 4\alpha\beta}}{2\beta} \right) . \qquad ∎$$

For $a \in \mathbb{N}$, let

$$\mathcal{H}_a = \left\{ (x, y) \in \Omega : \frac{1}{a+1} < y \leq \frac{1}{a} \right\}$$

$$\mathcal{V}_a = \left\{ (x, y) \in \Omega : \frac{1}{a+1} \leq x \leq \frac{1}{a} \right\}, \quad \text{for } a \geq 2, \qquad \text{and}$$

$$\mathcal{V}_1 = \left\{ (x, y) \in \Omega : \frac{1}{2} < x < 1 \right\} .$$

Notice that $T\mathcal{V}_a = \mathcal{H}_a$, and that

$$T^n(x, y) \in \mathcal{V}_a \Leftrightarrow a_{n+1} = a, \ n \geq 0,$$

$$T^n(x, y) \in \mathcal{H}_a \Leftrightarrow a_n = a, \ n \geq 1,$$

where $x = [0; a_1, a_2, \dots]$.

Exercise 5.3.4. Let $\mathcal{V}_a^* := \psi(\mathcal{V}_a)$ and $\mathcal{H}_a^* := \psi(\mathcal{H}_a)$, where $a \geq 1$. Show for $a \geq 2$ that $\mathcal{V}_a^*$ is a quadrangle with vertices

$$\left(0, \frac{1}{a} \right), \quad \left(\frac{a}{a+1}, \frac{1}{a+1} \right), \quad \left(\frac{a+1}{a+2}, \frac{1}{a+2} \right), \quad \text{and} \quad \left(0, \frac{1}{a+1} \right),$$

and that $\mathcal{H}_a^*$ is the reflection of $\mathcal{V}_a^*$ through the diagonal $\alpha = \beta$. For $a = 1$ both quadrangles reduce to triangles.   ∎

Exercise 5.3.5. Define the operator $K : \Gamma \to \Gamma$ by setting $K := \psi T \psi^{-1}$, and use Exercise 5.3.3 to show that

$$(\alpha, \beta) \in \mathcal{V}_a^* \to K(\alpha, \beta) = (\beta, \alpha + a\sqrt{1 - 4\alpha\beta} - a^2\beta).$$

and

$$(\alpha, \beta) \in \mathcal{H}_a^* \to K^{-1}(\alpha, \beta) = (\beta + a\sqrt{1 - 4\alpha\beta} - a^2\alpha, \alpha). \quad \blacksquare$$

Since

$$\psi(T_n, V_n) = (\Theta_{n-1}, \Theta_n),$$

and

$$(\Theta_n, \Theta_{n+1}) = \psi(T(\psi^{-1}(\Theta_{n-1}, \Theta_n))),$$

the following proposition is a direct corollary of Exercise 5.3.5.

Proposition 5.3.6. *Let x be an irrational number, with continued fraction expansion $x = [0; a_1, a_2, \dots]$. Then*

$$\Theta_{n+1} = \Theta_{n-1} + a_{n+1}\sqrt{1 - 4\Theta_{n-1}\Theta_n} - a_{n+1}^2\Theta_n, \quad n \geq 0,$$

and

$$\Theta_{n-1} = \Theta_{n+1} + a_{n+1}\sqrt{1 - 4\Theta_n\Theta_{n+1}} - a_{n+1}^2\Theta_n, \quad n \geq 1.$$

Exercise 5.3.7. For $a \geq 1$, let ξ_a be the irrational number given by

$$\xi_a = [0; \bar{a}],$$

i.e., ξ_a has a periodic continued fraction expansion with period length 1. Show that

$$\xi_a = \frac{-a + \sqrt{a^2 + 4}}{2}, \quad T(\xi_a, \xi_a) = (\xi_a, \xi_a),$$

and

$$\psi(\xi_a, \xi_a) = \left(\frac{1}{\sqrt{a^2+4}}, \frac{1}{\sqrt{a^2+4}} \right) . \qquad \blacksquare$$

It follows from Exercises 5.3.3, 5.3.5 and 5.3.7, that

$$\left(\frac{1}{\sqrt{a^2+4}}, \frac{1}{\sqrt{a^2+4}} \right) = K \left(\frac{1}{\sqrt{a^2+4}}, \frac{1}{\sqrt{a^2+4}} \right) , \quad \text{for } a \geq 1.$$

In view of this we define for $a \geq 1$ the map $f_a : \mathcal{V}_a^* \to \mathbb{R}$ by

$$f_a(\alpha, \beta) := \alpha + a\sqrt{1 - 4\alpha\beta} - a^2\beta .$$

Exercise 5.3.8. Show that for any (α, β) in the interior of $\mathcal{V}_a^*$ one has

$$\frac{\partial}{\partial \alpha} f_a(\alpha, \beta) < 0 \quad \text{and} \quad \frac{\partial}{\partial \beta} f_a(\alpha, \beta) < 0 . \qquad \blacksquare$$

Exercise 5.3.9. Use

$$f_a \left(\frac{1}{\sqrt{a^2+4}}, \frac{1}{\sqrt{a^2+4}} \right) = \frac{1}{\sqrt{a^2+4}} ,$$

and Exercise 5.3.7 to finish the proof of Theorem 5.3.1. $\qquad \blacksquare$

5.3.2 Metrical properties

In the early 1980s H.W. Lenstra formulated the following important conjecture on the distribution of the approximation coefficients $\Theta_n(x)$. This conjecture had previously been formulated (in a slightly different way) by W. Doeblin [Doe40], but had been completely forgotten. Our discussion requires the following definition: a function $F : \mathbb{R} \to [0, 1]$ is a *distribution function* if it is a non-decreasing function, that is right continuous and that satisfies $\lim_{x \to -\infty} F(x) = 0$ and $\lim_{x \to -\infty} F(x) = 1$.

For almost all x and for all $z \in [0, 1]$ the limit

$$\lim_{N \to \infty} \frac{1}{N} \#\{1 \leq n \leq N; \Theta_n(x) \leq z\}$$

exists, and equals the (distribution) function $F(z)$, given by

$$F(z) = \begin{cases} \dfrac{z}{\log 2}, & 0 \le z \le \frac{1}{2}, \\ \dfrac{1}{\log 2}(1 - z + \log 2z), & \frac{1}{2} < z \le 1. \end{cases}$$

In other words: For almost all x the sequence $(\Theta_n(x))_{n \ge 1}$ has limiting distribution F. Notice that, once this conjecture is established, we have—simply by calculating the expectation of F—that for almost all x

$$\lim_{N \to \infty} \frac{1}{N} \sum_{j \le N} \Theta_j(x) = \frac{1}{4 \log 2} = 0.360673 \ldots .$$

The Doeblin-Lenstra conjecture was proved by W. Bosma, H. Jager and F. Wiedijk [BJW83]. Here we will present a proof due to Jager [Jag86].

When one tries to prove this conjecture by applying the ergodic-theoretic apparatus, which seems only natural, one soon realizes that the ergodic system $([0, 1), \mathcal{L}, \mu, T)$ underlying the (regular) continued fraction is insufficient. From (5.9) it is obvious that $\Theta_n(x)$ depends on both the future (i.e., $T_n = T^n(x)$) and the past (i.e., V_n) of x; T forgets about the past (the same applies also to $\Theta_{n-1}(x)$). It is for this reason Bosma, Jager and Wiedijk turned to the natural extension $(\Omega, \mathcal{L}, \bar{\mu}, \mathcal{T})$ of $([0, 1), \mathcal{L}, \mu, T)$; they wanted to have a grip on the past.

One problem in studying the dynamic behaviour of the orbit $\mathcal{T}^n(x, 0) = (T_n, V_n), n \ge 0$, is that the collection of such orbits as x varies in $[0, 1)$—which is $N = [0, 1) \times ([0, 1] \cap \mathbb{Q})$—has measure zero with respect to the measure $\bar{\mu}$. Due to the Ergodic Theorem we know that for almost every $(x, y) \in [0, 1) \times [0, 1]$, and for any Borel set C of positive measure,

$$\lim_{N \to \infty} \frac{1}{N} \sum_{k=0}^{N-1} 1_C \left(\mathcal{T}^k(x, y) \right) = \bar{\mu}(C),$$

Definition 5.3.10. *Let* $((x_n, y_n))_{n \ge 0}$ *be a sequence in* $[0, 1) \times [0, 1]$. *Then the sequence* $((x_n, y_n))_{n \ge 0}$ *is distributed over* $[0, 1) \times [0, 1]$ *with* *density function* $d(x, y) = (\log 2)^{-1}(1 + xy)^{-2}$, *if*

$$\lim_{N \to \infty} \frac{1}{N} \sum_{k=0}^{N-1} 1_J(x_k, y_k) = \bar{\mu}(J)$$

holds for all rectangles $J = \langle a, b \rangle \times \langle c, d \rangle \subset [0, 1) \times [0, 1]$. *Here* $0 \leq a < b < 1$, $0 \leq c < d \leq 1$, *and* $\langle a, b \rangle$ *is one of the intervals* $[a, b]$, $[a, b)$, $(a, b]$ *or* (a, b).

This definition is a two-dimensional generalization of the classical definition of a *uniformly distributed sequence*; see e.g. [KN74] and [Hla84]. Clearly, if $((x_n, y_n))_{n \geq 0}$ is distributed over $[0, 1) \times [0, 1]$ with density function d, then one also has that

$$\lim_{N \to \infty} \frac{1}{N} \sum_{k=0}^{N-1} 1_C(x_k, y_k) = \frac{1}{\log 2} \iint_C \frac{dxdy}{(1 + xy)^2} = \bar{\mu}(C) \quad (5.14)$$

for any set C that is the countable union of (complements of) rectangles J. In fact one can show that (5.14) holds if $C \in \mathcal{B}$ is such that $\bar{\mu}(\partial C) = 0$; see [KN74], pp. 174, 175.

The following lemma is very useful.

Lemma 5.3.11. (Jager, 1986) *For almost all irrational numbers* $x \in [0, 1)$ *the two-dimensional sequence*

$$(T^n(x, 0))_{n \geq 0} = (T_n, V_n)_{n \geq 0}$$

is distributed over $[0, 1) \times [0, 1]$ *with density function* $d(x, y) = (\log 2)^{-1}(1 + xy)^{-2}$.

Proof. Let $\varepsilon > 0$ be arbitrary. For a rectangle $J = \langle a, b \rangle \times \langle c, d \rangle$, define sets $J_{+\varepsilon}$ and $J_{-\varepsilon}$ by

$$J_{+\varepsilon} = \langle a, b \rangle \times [c - \varepsilon, d + \varepsilon] \quad \text{and} \quad J_{-\varepsilon} = \langle a, b \rangle \times [c + \varepsilon, d - \varepsilon].$$

For any $(x, y) \in ([0, 1) \setminus \mathbb{Q}) \times [0, 1]$, with $x = [0; a_1, \ldots]$, there exists a uniform $n_0 = n_0(\varepsilon) \geq 1$ such that

$$|[0; a_n, a_{n-1}, \ldots, a_1 + y] - [0; a_n, a_{n-1}, \ldots, a_1]| < \varepsilon$$

for all $n \geq n_0$. But then we have for any $n \geq n_0$

$$T^n(x, y) = (T_n, [0; a_n, a_{n-1}, \ldots, a_1 + y]) \in J_{-\varepsilon} \Rightarrow (T_n, V_n) \in J$$

and

$$(T_n, V_n) \in J \quad \Rightarrow \quad T^n(x, y) \in J_{+\varepsilon},$$

which yields that

$$\liminf_{N \to \infty} \frac{1}{N} \sum_{k=0}^{N-1} 1_{J_{-\varepsilon}} \left(T^k(x, y) \right) \leq \liminf_{N \to \infty} \frac{1}{N} \sum_{k=0}^{N-1} 1_J \left(T^k(x, 0) \right)$$

$$\leq \limsup_{N \to \infty} \frac{1}{N} \sum_{k=0}^{N-1} 1_J \left(T^k(x, 0) \right)$$

$$\leq \limsup_{N \to \infty} \frac{1}{N} \sum_{k=0}^{N-1} 1_{J_{+\varepsilon}} \left(T^k(x, y) \right).$$

From this and the Ergodic Theorem we find that for almost all x

$$\bar{\mu}(J_{-\varepsilon}) \leq \liminf_{N \to \infty} \frac{1}{N} \sum_{k=0}^{N-1} 1_{J_{-\varepsilon}} \left(T^k(x, 0) \right)$$

$$\leq \limsup_{N \to \infty} \frac{1}{N} \sum_{k=0}^{N-1} 1_J \left(T^k(x, 0) \right) \leq \bar{\mu}(J_{+\varepsilon}).$$

Since

$$|\bar{\mu}(J_{\pm\varepsilon}) - \bar{\mu}(J)| \leq \frac{\varepsilon}{\log 2},$$

it follows that

$$\bar{\mu}(J) - \frac{\varepsilon}{\log 2} \leq \liminf_{N \to \infty} \frac{1}{N} \sum_{k=0}^{N-1} 1_{J_{-\varepsilon}} \left(T^k(x, 0) \right)$$

$$\leq \limsup_{N \to \infty} \frac{1}{N} \sum_{k=0}^{N-1} 1_J \left(T^k(x, 0) \right) \leq \bar{\mu}(J) + \frac{\varepsilon}{\log 2},$$

from which the conclusion of the lemma follows. ∎

Setting for $0 \leq z \leq 1$,

$$A_z := \left\{ (x, y) \in \Omega; \ \frac{x}{1 + xy} \leq z \right\},$$

it follows from Lemma 5.3.11 that for almost all x

$$\lim_{N \to \infty} \frac{1}{N} \sum_{i=0}^{N-1} 1_{A_z} \left(T^i(x, 0) \right) = \bar{\mu}(A_z).$$

Exercise 5.3.12. Show by direct calculation that $F(z) = \bar{\mu}(A_z)$. This proves the Doeblin-Lenstra conjecture. ∎

In view of (5.9) one can also look at the distribution of two consecutive Θ's.

Exercise 5.3.13. (Jager, 1986) Let $(z_1, z_2) \in \Gamma$. Show that for almost all x the limit

$$\lim_{N \to \infty} \frac{1}{N} \#\{1 \leq n \leq N; \ \Theta_{n-1}(x) \leq z_1, \ \Theta_n(x) \leq z_2\}$$

exists, and equals the (distribution) function $G(z_1, z_2)$, given by

$$G(z_1, z_2) = \frac{1}{\log 2} \left(1 - \sqrt{1 - 4z_1 z_2} + \log \frac{1}{2}(1 + \sqrt{1 - 4z_1 z_2}) \right). \ \blacksquare$$

In fact, we already saw that $(\Theta_{n-1}(x), \Theta_n(x))$ lives in its own space Γ. To be more precise, we have the following exercise.

Exercise 5.3.14. (Jager, 1986) Show that the dynamical system (Γ, ρ, K), with ρ the probability measure on Γ with density

$$\frac{1}{\log 2}\frac{1}{\sqrt{1 - 4\alpha\beta}},$$

forms an ergodic system.   ∎

Clearly (Γ, ρ, K) is better suited to derive the Doeblin-Lenstra conjecture and the result from Exercise 5.3.13. We decided to present this material in its historical order, so you would get a better understanding of the dynamics of the system $(\Omega, \mathcal{L}, \bar{\mu}, \mathcal{T})$.

5.4   Other continued fractions

5.4.1   Introduction

Apart from the regular continued fractions (RCF) an—at first sight—bewildering variety of continued fraction expansions exist. Here we mention the continued fraction to the nearer integer (NICF), Hurwitz' singular continued fraction, the continued fraction with odd (or even) partial quotients, Rosen's λ-expansions, Nakada's α-expansions, Bosma's optimal continued fraction, ... Enough to fill several pages with names of different continued fraction expansions.

Fortunately, many of these expansions are inter-related via inducing or via the integral transformation (or both), which we studied in the previous chapter. In this section we will describe a class of continued fraction expansions that can be obtained from the RCF via inducing; these are the *S-expansions*; see also [Kra91]. These *S*-expansions are examples of *semi-regular continued fraction expansions*, defined as follows.

Definition 5.4.1. *A semi-regular continued fraction* (SRCF) *is a finite or infinite fraction*

$$b_0 + \cfrac{\varepsilon_1}{b_1 + \cfrac{\varepsilon_2}{b_2 + \cfrac{\ddots}{\cdot} + \cfrac{\varepsilon_n}{b_n + \ddots}}}, \qquad (5.15)$$

with $\varepsilon_n = \pm 1$; $b_0 \in \mathbb{Z}$; $b_n \in \mathbb{N}$, for $n \geq 1$, subject to the condition

$$\varepsilon_{n+1} + b_n \geq 1, \ for \ n \geq 1,$$

and with the restriction that in the infinite case

$$\varepsilon_{n+1} + b_n \geq 2, \ infinitely \ often.$$

Moreover we demand that $\varepsilon_n + b_n \geq 1$ for $n \geq 1$. We will abbreviate the SRCF from (5.15) by

$$[b_0; \varepsilon_1 b_1, \varepsilon_2 b_2, \ldots, \varepsilon_n b_n, \ldots].$$

Example 5.4.2. One of the most important examples of a semi-regular continued fraction expansion is the *nearest integer continued fraction expansion* (NICF). Introduced in 1873 by Minnigerode [Min73] and studied by Hurwitz [Hur89], this SRCF derives its name from the fact that the partial quotients are always the nearest integers. To be less poetic and more precise, let $T_{\frac{1}{2}} : [-\frac{1}{2}, \frac{1}{2}) \to [-\frac{1}{2}, \frac{1}{2})$ be given by

$$T_{\frac{1}{2}}(x) := \left| \frac{1}{x} \right| - \left\lfloor \left| \frac{1}{x} \right| + \frac{1}{2} \right\rfloor, \quad x \neq 0; \quad T_{\frac{1}{2}}(0) := 0.$$

For $x \in \mathbb{R}$, setting $b_0 = b_0(x) \in \mathbb{Z}$ such that $x - b_0 \in [-\frac{1}{2}, \frac{1}{2})$, the partial quotients $b_n = b_n(x)$ and the 'signs' $\varepsilon_n = \varepsilon_n(x)$ are defined as follows. Let $n \geq 1$ be such that $T_{\frac{1}{2}}^{n-1}(x - b_0) \neq 0$; then

$$\varepsilon_n = \text{sgn} \left(T_{\frac{1}{2}}^{n-1}(x - b_0) \right),$$

where

$$\text{sgn}(x) = \begin{cases} 1, & \text{if } x > 0, \\ 0, & \text{if } x = 0, \\ -1, & \text{if } x < 0, \end{cases}$$

and

$$b_n := \left\lfloor \left| \frac{1}{T_{\frac{1}{2}}^{n-1}(x - b_0)} \right| + \frac{1}{2} \right\rfloor .$$

Since

$$x = b_0 + \frac{\varepsilon_1}{b_1 + T_{\frac{1}{2}}(x - b_0)}$$

we find that x has (5.15) as its NICF-expansion. In this case one moreover has that

$$b_n \geq 2 \quad \text{and} \quad b_n + \varepsilon_{n+1} \geq 2, \text{ for } n \geq 1. \qquad (5.16)$$

Conversely, if x has an SRCF-expansion of the form (5.15) which also satisfies (5.16), then this SRCF-expansion (5.15) is the NICF-expansion of x.   ∎

Exercise 5.4.3. Let $x = [0; \overline{1, 1, 2}]$.

(a) Show that $x = -1 + \frac{1}{2}\sqrt{10} = 0.5811\ldots$.

(b) Determine the first 10 RCF-convergents of x. (*Hint:* Use the recurrence relations from Exercise 1.3.8.)

(c) Determine the NICF-expansion of x, and calculate via finite truncation the first six NICF-convergents of x. Compare these with the RCF-convergents of x. What do you see?   ∎

Let $x \in \mathbb{R}$, and suppose that (5.15) is some SRCF-expansion of x. Finite truncation in (5.15) yields the convergents $(r_k/s_k)_{k \geq 1}$ of x of this particular SRCF. In a moment we will see that, for instance, the sequence of NICF-convergents $(r_k/s_k)_{k \geq 1}$ of x forms a subsequence of the RCF-convergents $(p_n/q_n)_{n\geq 1}$ of x; i.e., there exists a function $n_x = n : \mathbb{N} \to \mathbb{N}$ such that

$$\frac{r_k}{s_k} = \frac{p_{n(k)}}{q_{n(k)}}, \quad k \geq 1.$$

We will also see that for almost all x

$$\lim_{k \to \infty} \frac{n(k)}{k} = \frac{\log 2}{\log G} = 1.4404 \ldots,$$

with G the golden mean from Example 1.3.4, so the NICF-convergents provide a faster approximation to x than the RCF-convergents. Furthermore we will see, setting

$$\theta_k = \theta_k(x) := s_k^2 \left| x - \frac{r_k}{s_k} \right|, \quad k \geq 1,$$

that for almost all x one has

$$\lim_{k \to \infty} \frac{1}{k} \sum_{i=1}^{k} \theta_i = \frac{\sqrt{5} - 2}{2 \log G} = 0.24528 \ldots,$$

so the NICF also gives—in the mean—closer approximations to x than the RCF does. This raises the natural question whether there exist SRCF-expansions of x that are faster and closer to x than the NICF. We will see that there exists a unique SRCF-expansion, the *optimal continued fraction expansion* (OCF), which is as fast as the NICF, but which yields the closest possible approximations to *any* irrational x.

Example 5.4.4. In 1981, H. Nakada [Nak81] generalized the idea behind the NICF in the following way. Let $\alpha \in [\frac{1}{2}, 1]$, and let the map $T_\alpha : [\alpha - 1, \alpha) \to [\alpha - 1, \alpha)$ be given by

$$T_\alpha(x) := \left| \frac{1}{x} \right| - \left\lfloor \left| \frac{1}{x} \right| + 1 - \alpha \right\rfloor, \quad x \neq 0; \quad T_\alpha(0) := 0.$$

As in the case of the NICF, the map T_α generates a semi-regular continued fraction expansion, the *α-expansion*. For $x \in \mathbb{R}$, setting $b_0 = b_0(x) \in \mathbb{Z}$ such that $x - b_0 \in [\alpha - 1, \alpha)$, the partial quotients

$b_n = b_n(x)$ and the signs $\varepsilon_n = \varepsilon_n(x)$ of the α-expansion of x are defined as follows. Let $n \geq 1$ be such that $T_\alpha(x - b_0) \neq 0$; then

$$\varepsilon_n := \operatorname{sgn}(T_\alpha^{n-1}(x - b_0))$$

and

$$b_n := \left\lfloor \left| \frac{1}{T_\alpha^{n-1}(x - b_0)} \right| + 1 - \alpha \right\rfloor.$$

Since

$$x = b_0 + \frac{\varepsilon_1}{b_1 + T_\alpha(x - b_0)},$$

we find that x has (5.15) as its α-expansion. Obviously the case $\alpha = \frac{1}{2}$ gives the NICF, while $\alpha = 1$ yields the RCF. In the next exercise we consider the case $\alpha = g$, the "small" golden mean. In [Nak81] Nakada gives the natural extension for each α-expansion. In the next subsection we will re-obtain—albeit in a completely different way—his results.

■

Exercise 5.4.5. A continued fraction closely related to the NICF is *Hurwitz' singular continued fraction* (SCF). The operator underlying the SCF is the map $T_g : [g - 1, g) \to [g - 1, g)$, given by

$$T_g(x) := \left| \frac{1}{x} \right| - \left\lfloor \left| \frac{1}{x} \right| + 1 - g \right\rfloor, \quad x \neq 0; \quad T_g(0) := 0.$$

(a) Let x be an irrational number, with SCF-expansion (5.15). Show that

$$b_n \geq 2 \quad \text{and} \quad b_n + \varepsilon_n \geq 2, \quad \text{for } n \geq 1. \qquad (5.17)$$

(Further on in this section we will see that also the converse holds, i.e., if x has an SRCF-expansion of the form (5.15) which also satisfies (5.17), then this SRCF-expansion (5.15) is the SCF-expansion of x.)

(b) Let $x = [0; \overline{1, 1, 2}]$. Determine the SCF-expansion of x, and calculate the first six SCF-convergents of x. Compare these with the RCF- and NICF-convergents of x you found in Exercise 5.4.3. What do you see?   ∎

Example 5.4.6. Let (5.15) be some SRCF-expansion of $x \in \mathbb{R}$, with convergents $(r_k/s_k)_{k \geq 1}$ and approximation coefficients $(\theta_k)_{k \geq 1}$. This SRCF-expansion is called Minkowski's *diagonal continued fraction expansion* (DCF) of x if it satisfies $\theta_k < \frac{1}{2}$, for $k \geq 1$. The DCF-expansion of any $x \in \mathbb{R}$ is unique. In the next subsection we will see that it can be obtained in a very simple way from the RCF-expansion of x via singularizations.   ∎

5.4.2 S-expansions

At first one might think there is no relation whatsoever between the semiregular continued fraction expansions we have just discussed. However, there is a simple—and quite old—procedure, called *singularization*, which links these (and very many other SRCF-expansions) to the RCF expansion. A singularization is based on a simple algebraic manipulation on partial quotients of SRCF-expansions, described in the following exercise.

Exercise 5.4.7. Let a, b be positive integers, and let $\xi \in [0, 1)$. Show that the following identity holds:

$$a + \cfrac{1}{1 + \cfrac{1}{b + \xi}} = a + 1 + \frac{-1}{b + 1 + \xi} .$$ ∎

We now define the notion of singularization based on this idea. Let x be an irrational number, and let (5.15) be some SRCF expansion of x with sequence of convergents $(r_k/s_k)_{k \geq 1}$, where

$$\frac{r_k}{s_k} = [b_0; \varepsilon_1 b_1, \varepsilon_2 b_2, \ldots, \varepsilon_k b_k]$$

and $(r_k, s_k) = 1$, $s_k > 0$. Suppose that for a certain $\ell \geq 0$ we have that $b_{\ell+1} = 1$ and that $\varepsilon_{\ell+1} = \varepsilon_{\ell+2} = 1$. The operation by which the continued fraction (5.15) is replaced by

$$[b_0; \varepsilon_1 b_1, \ldots, \varepsilon_{\ell-1} b_{\ell-1}, \varepsilon_\ell(b_\ell + 1), -(b_{\ell+2} + 1),$$
$$\varepsilon_{\ell+3} b_{\ell+3}, \varepsilon_{\ell+4} b_{\ell+4}, \ldots],$$

which again is an SRCF-expansion of x, with convergents, say, $(c_n/d_n)_{n \geq -1}$, is called the *singularization of the partial quotient* $b_{\ell+1}$ *equal to* 1. (In case $\ell = 0$ this comes down to replacing (5.15) by $[b_0 + 1; -(b_2 + 1), \varepsilon_3 b_3, \varepsilon_4 b_4, \ldots]$.) One easily shows that $(c_n/d_n)_{n \geq -1}$ is obtained from $(r_k/s_k)_{k \geq -1}$ by skipping the term r_ℓ/s_ℓ. See also [Kra91], Sections 2 and 4.

Exercise 5.4.8. Let $x = [0; \overline{1, 1, 2}]$; see also Exercise 5.4.7.

(i) Singularize the first partial quotient of x equal to 1 (this is a_1), and repeat this 5 times recursively. What are the convergents of this 'new' SRCF expansion of x? Could you have done the singularization in 'one stroke'? Which expansion did you obtain?

(ii) Singularize the second partial quotient of x equal to 1 (this is a_2), and repeat this 5 times recursively. What are the convergents of this 'new' SRCF expansion of x? Again, could you have done these singularizations in 'one stroke'? Which expansion did you obtain?

∎

The following two exercises 'generalize' the previous exercise.

Exercise 5.4.9. Let $x \in [0, 1)$ be some irrational number, with RCF-expansion $x = [0; a_1, a_2, \ldots]$. Now consider the following algorithm.

Singularize in each block of m consecutive partial quotients

$$a_{n+1} = 1, \ldots, a_{n+m} = 1,$$

where $m \in \mathbb{N} \cup \{\infty\}$, $a_{n+m+1} \neq 1$ and $a_n \neq 1$ in case $n > 0$, the first, third, fifth, etc. partial quotient.

After applying this algorithm to the RCF-expansion of x one obtains an SRCF-expansion $[b_0; \varepsilon_1 b_1, \varepsilon_2 b_2, \ldots, \varepsilon_n b_n, \ldots]$ of x. Show this is the NICF-expansion of x. (*Hint:* Show that (5.16) is satisfied.) ∎

Exercise 5.4.10. Again, let $x \in [0, 1)$ be some irrational number, with RCF-expansion $x = [0; a_1, a_2, \ldots]$. Now consider the following variation on the algorithm from Exercise 5.4.9.

Singularize in each block of m consecutive partial quotients

$$a_{n+1} = 1, \ldots, a_{n+m} = 1,$$

where $m \in \mathbb{N}$ is odd, $a_{n+m+1} \neq 1$ and $a_n \neq 1$ in case $n > 0$, or in case $m = \infty$

singularize the first, third, fifth, etc. partial quotient.

In case $m \in \mathbb{N}$ is even

singularize the second, fourth, sixth, etc. partial quotient.

After applying this algorithm to the RCF-expansion of x one obtains an SRCF-expansion $[b_0; \varepsilon_1 b_1, \varepsilon_2 b_2, \ldots, \varepsilon_n b_n, \ldots]$ of x. Show this is the SCF-expansion of x. (*Hint:* Show that (5.17) is satisfied.) ∎

A simple way to derive a strategy for singularization is given by a *singularization area* S. Let $(\Omega, \overline{\mathcal{B}}, \bar{\mu}, \mathcal{T})$ be the natural extension of the RCF. Here we will choose S to be a subset of Ω. Before we give the definition we have two exercises, in order to introduce the idea in a natural (!) way.

Exercise 5.4.11. Let $x \in [0, 1)$ be some irrational number, with RCF-expansion $x = [0; a_1, a_2, \ldots]$. As before, let $T_n = T^n(x) =$

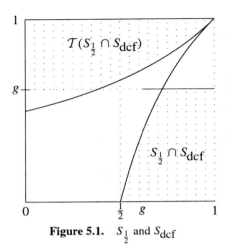

Figure 5.1. $S_{\frac{1}{2}}$ and S_{dcf}

$[0; a_{n+1}, a_{n+2}, \ldots]$ and $V_n = q_{n-1}/q_n = [0; a_n, a_{n-1}, \ldots, a_1]$, $n \geq 1$. Finally, let $S_{\frac{1}{2}} := [\frac{1}{2}, g) \times [0, g] \cup [g, 1) \times [0, g)$, where g is the "small" golden mean; see also Figure 5.1. Consider the following algorithm:

    Singularize $a_{n+1} = 1$ if and only if $(T_n, V_n) \in S_{\frac{1}{2}}$.

(a) Show that you never have to singularize two consecutive partial quotients equal to 1.

(b) After applying this algorithm to the RCF-expansion of x one obtains a SRCF-expansion $[b_0; \varepsilon_1 b_1, \varepsilon_2 b_2, \ldots, \varepsilon_n b_n, \ldots]$ of x. Show this is the NICF-expansion of x.   ■

Exercise 5.4.12. Let $x \in [0, 1)$ be some irrational number, with RCF-expansion $x = [0; a_1, a_2, \ldots]$. As before, let $T_n = T^n(x) = [0; a_{n+1}, a_{n+2}, \ldots]$ and $V_n = q_{n-1}/q_n = [0; a_n, a_{n-1}, \ldots, a_1]$, $n \geq 1$. Finally, let $S_g := [g, 1) \times [0, g) \cup (g, 1) \times [g, 1]$; see also Figure 5.1. Consider the following algorithm.

    Singularize $a_{n+1} = 1$ if and only if $(T_n, V_n) \in S_g$.

(a) Show that you never have to singularize two consecutive partial quotients equal to 1.

(b) After applying this algorithm to the RCF-expansion of x one obtains a SRCF-expansion $[b_0; \varepsilon_1 b_1, \varepsilon_2 b_2, \ldots, \varepsilon_n b_n, \ldots]$ of x. Show this is the SCF-expansion of x. ∎

The sets $S_{\frac{1}{2}}$ and S_g from Exercises 5.4.11 and 5.4.12 are examples of a so-called *singularization area*. Here is the formal definition.

Definition 5.4.13. *A subset S from Ω is called a singularization area if it satisfies*

(i) $S \subset [\frac{1}{2}, 1) \times [0, 1]$;

(ii) $T(S)$ *and S are either disjoint or intersect in the single point* (g, g);

(iii) $S \in \mathcal{B}$ *and* $\mu(\partial S) = 0$;

This definition reflects that we only singularize partial quotients equal to 1 (i), and that we never singularize two consecutive partial quotients equal to 1 (ii). Note that (g, g) is a fixed point of T. That S should be a Borel set is obvious; otherwise we would not be able to use ergodic theory. At first the requirement that $\mu(\partial S) = 0$ might be a bit mysterious. This is needed to exclude pathological cases. Consider for instance the set

$$S = \{(x, y) \in S_{\frac{1}{2}}; \ y \notin \mathbb{Q}\};$$

so S is $S_{\frac{1}{2}}$ with all its points (x, y) with rational second coordinate removed. Clearly this set S has the same measure as $S_{\frac{1}{2}}$, in fact $\partial S = S_{\frac{1}{2}}$, and (i) and (ii) from Definition 5.4.13 are satisfied. However, since (T_n, V_n) always has a rational second coordinate, no partial quotient is singularized.

Exercise 5.4.14. Use Definition 5.4.13 and Figure 5.1 to show that

$$0 \le \bar{\mu}(S) \le 1 - \frac{\log G}{\log 2} = 0.3057\ldots,$$

where $G = \frac{1}{2}(\sqrt{5} + 1)$. See also [Kra91], Theorem (4.7).   ∎

A singularization area is called *maximal* in case

$$\bar{\mu}(S) = 1 - \frac{\log G}{\log 2} = 0.3057\ldots.$$

Definition 5.4.15. *Let S be a singularization area and let $x \in [0, 1)$ be an irrational number. The S-expansion of x is the semi-regular continued fraction expansion converging to x that is obtained from the RCF-expansion $x = [0; a_1, a_2, \ldots]$ of x by singularizing a_{n+1} if and only if $T^n(x, 0) \in S, n \ge 0$.*

Exercise 5.4.16.

(a) Show that $S_{\frac{1}{2}}$ and S_g are maximal singularization areas; see [Kra91].

(b) Show that $S_{\text{ocf}} := \{(T, V) \in \Omega; V < \min(T, \frac{2T-1}{1-T})\}$ is a maximal singularization area. The S-expansion associated with this singularization is called the *optimal continued fraction* (OCF); see [BK90] and [BK91].

(c) Show that $S_{\text{dcf}} := \{(T, V) \in \Omega; \frac{T}{1+TV} > \frac{1}{2}\}$ is a non-maximal singularization area that yields the *diagonal continued fraction* (DCF) of Minkowski; see [Kra91].   ∎

Remark 5.4.17. That the NICF, SCF and OCF algorithms singularize blocks of odd length in the same way reflects the fact that these expansions are maximal. There is only one way to throw out (= to singularize) as many 1's as possible in a block of odd length. In a block of even length a jump has to be made somewhere; see also [Kra91]. E.g., for the NICF one makes this jump at the end, and for the SCF at the beginning. We will see that the OCF chooses the jump in such a way that one is left with the smallest possible θ_k's. One can show (see [BK90]) that for the OCF the jump takes place in the middle of the block.

That for a maximal S-expansion one always makes the maximal number of throw-outs in any block of consecutive 1's has several nice consequences. E.g., maximal S-expansions are isomorphic. An explicit isomorphism can be found in [Kra93]. ∎

Let S be a singularization area and let $x \in [0, 1)$ be a real irrational number, with RCF-expansion $x = [0; a_1, a_2, \dots]$ and RCF-convergents $(p_n/q_n)_{n \geq 1}$. Furthermore, let $[b_0; \varepsilon_1 b_1, \dots, \varepsilon_k b_k, \dots]$ be the S-expansion of x, and let r_k/s_k, $k \geq 1$ be its S-convergents, i.e.,

$$\frac{r_k}{s_k} := [b_0; \varepsilon_1 b_1, \dots, \varepsilon_k b_k], \quad k \geq 1.$$

Due to the singularization mechanism one obviously has that $(r_k/s_k)_{k \geq 1}$ forms a subsequence of $(p_n/q_n)_{n \geq 1}$. But then there exists a monotone function $n_S : \mathbb{N} \to \mathbb{N}$ such that

$$\begin{pmatrix} r_k \\ s_k \end{pmatrix} = \begin{pmatrix} p_{n_S(k)} \\ q_{n_S(k)} \end{pmatrix}, \quad k \geq 1.$$

We have the following proposition, which can be seen as a special case of *Kac's Lemma*. See Exercise 4.2.5(c).

Proposition 5.4.18. ([Kra91]) *Let S be a singularization area. Then for almost all x one has*

$$\lim_{k \to \infty} \frac{n_S(k)}{k} = \frac{1}{1 - \bar{\mu}(S)}.$$

Proof. From the definition of n_S it follows that

$$n_S(k) = k + \sum_{j=1}^{n_S(k)} 1_S(T_j, V_j).$$

It follows from Lemma 5.3.11 that

$$\lim_{k \to \infty} \frac{1}{n_S(k)} \sum_{j=1}^{n_S(k)} 1_S(T_j, V_j) = \bar{\mu}(S),$$

and the result follows from

$$1 = \frac{k}{n_S(k)} + \frac{1}{n_S(k)} \sum_{j=1}^{n_S(k)} 1_S(T_j, V_j). \qquad \blacksquare$$

Remark 5.4.19. Notice that Proposition 5.4.18 implies that

$$1 \le \lim_{k \to \infty} \frac{n_S(k)}{k} \le \frac{\log 2}{\log G} = 1.4404 \cdots,$$

the upper bound being attained if and only if S is maximal. Since we saw that S is maximal for the nearest integer continued fraction expansion, we have obtained a theorem of William W. Adams [Ada79]; see also [Jag82] for a third proof. In fact many results previously obtained for the nearest integer continued fraction now hold generally for any S-expansion. $\qquad \blacksquare$

We have the following corollary.

Corollary 5.4.20. *Let S be a singularization area and let x be an irrational number with S-expansion $[b_0; \varepsilon_1 b_1, \dots]$ and S-convergents r_k/s_k, $k \ge 1$. Then for almost all x one has*

$$\lim_{k \to \infty} \frac{1}{k} \log s_k = \frac{1}{1 - \bar{\mu}(S)} \frac{\pi^2}{12 \log 2},$$

and

$$\lim_{k \to \infty} \frac{1}{k} \log \left| x - \frac{r_k}{s_k} \right| = \frac{1}{1 - \bar{\mu}(S)} \frac{-\pi^2}{6 \log 2}.$$

Proof. For the first statement, notice that

$$\frac{1}{k} \log s_k = \frac{n_S(k)}{k} \frac{1}{n_S(k)} \log q_{n_S(k)}.$$

But then the first statement follows for almost all x from Lévy's Proposition 3.5.5 and Proposition 5.4.18; see also Equation (3.7). In the same way the second statement follows from Equation (3.9). ∎

For any irrational number x with RCF-expansion $[a_0; a_1, \dots]$ and S-expansion $[b_0; \varepsilon_1 b_1, \dots]$, we define the shift t by

$$t(x - b_0) := [0; \varepsilon_2 b_2, \dots, \varepsilon_k b_k, \dots].$$

For a fixed x and for $k \geq 0$ we put

$$t_k := t^k(x - b_0) = [0; \varepsilon_{k+1} b_{k+1}, \varepsilon_{k+2} b_{k+2}, \dots] \quad \text{and} \quad v_k := s_{k-1}/s_k,$$

where

$$v_k = [0; b_k, \varepsilon_k b_{k-1}, \dots, \varepsilon_2 b_1], \; k \geq 1; \quad v_0 = 0.$$

See also [Kra91], (1.4) and (5.1). We have the following theorem, which is given here without proof.

Theorem 5.4.21. ([Kra91]) *Let S be a singularization area and put $\Delta_S := \Omega \setminus S$, $\Delta_S^- := \mathcal{T} S$ and $\Delta_S^+ := \Delta_S \setminus \Delta_S^-$. Let x be a real number, with RCF-expansion $[a_0; a_1, \dots]$ and RCF-convergents $(p_n/q_n)_{n \geq 1}$. Then one has:*

1. *The system $(\Delta_S, \mathcal{L}, \rho_S, \mathcal{O}_S)$ forms an ergodic system. Here ρ_S is the probability measure on $(\Delta_S, \mathcal{L})$ with density*

$$((1 - \mu(S)) \log 2)^{-1} (1 + xy)^{-2}$$

and the map $\mathcal{O}_S$ is induced by $\mathcal{T}$ on Δ_S, i.e.,

$$\mathcal{O}_S(x, y) = \begin{cases} \mathcal{T}(x, y), & \mathcal{T}(x, y) \in \Delta_S, \\ \mathcal{T}^2(x, y), & \mathcal{T}(x, y) \in S; \end{cases}$$

2. $\mathcal{T}^n(x, 0) \in S \iff p_n/q_n$ *is not an S-convergent;*

3. p_n/q_n *is not an S-convergent $\implies$ both p_{n-1}/q_{n-1} and p_{n+1}/q_{n+1} are S-convergents;*

4. $T^n(x,0) \in \Delta_S^+ \Leftrightarrow \exists k :$
$$\begin{cases} r_{k-1} = p_{n-1}, & r_k = p_n \\ s_{k-1} = q_{n-1}, & s_k = q_n \end{cases}$$
$$\text{and } T^n(x,0) = (t_k, v_k) ;$$

5. $T^n(x,0) \in \Delta_S^- \Leftrightarrow \exists k :$
$$\begin{cases} r_{k-1} = p_{n-2}, & r_k = p_n \\ s_{k-1} = q_{n-2}, & s_k = q_n \end{cases}$$
$$\text{and } T^n(x,0) = (\tfrac{-t_k}{1+t_k}, 1 - v_k) .$$

(See also [Kra91], Theorem (5.3).)

The following exercise is a direct consequence of the Ergodic Theorem and the fact that the system $(\Delta_S, \mathcal{L}, \rho_S, \mathcal{O}_S)$ forms an ergodic system.

Exercise 5.4.22. Let S be a singularization area and let x be an irrational number with S-expansion $[b_0; \varepsilon_1 b_1, \dots]$ and S-convergents $r_k/s_k, k \geq 1$. Show that for almost all x one has

$$\lim_{k \to \infty} \frac{1}{k} \sum_{i=1}^{k} \varepsilon_i = \frac{1 - 3\bar{\mu}(S)}{1 - \bar{\mu}(S)} . \qquad \blacksquare$$

In view of Theorem 5.4.21 we define the map $\mathcal{M} : \Delta_S \to \mathbb{R}^2$ by

$$\mathcal{M}(T, V) := \begin{cases} (T, V), & (T, V) \in \Delta_S^+, \\ \left(\dfrac{-T}{1+T}, 1 - V \right), & (T, V) \in \Delta_S^- . \end{cases}$$

We have the following theorem.

Theorem 5.4.23. ([Kra91]) *Let S be a singularization area and put $\Omega_S := \mathcal{M}(\Delta_S)$. Let $\mathcal{L}$ be the collection of Lebesgue subsets of Ω_S and let μ_S be the probability measure on $(\Omega_S, \mathcal{L})$, defined by*

$$\mu_S(E) := \rho_S(\mathcal{M}^{-1}(E)), \ E \in \mathcal{L} .$$

If we also define the map $T_S : \Omega_S \to \Omega_S$ by

$$T_S(t, v) := \mathcal{M}(\mathcal{O}_S(\mathcal{M}^{-1}(t, v))), \quad (t, v) \in \Omega_S,$$

then T_S is isomorphic to $\mathcal{O}_S$ by $\mathcal{M}$, and $(\Omega_S, \mathcal{L}, \mu_S, T_S)$ forms an ergodic system with density $((1 - \mu(S))\log 2)^{-1}(1 + tv)^{-2}$. Finally, for almost all $x \in [0, 1)$ the sequence $(t_k, v_k)_{k \geq 0}$ is distributed over Ω_S according to this density.

Exercise 5.4.24. Let $S = S_{\frac{1}{2}}$, the singularization area of the NICF.

(a) Show that Ω_S is given by $\Delta^+ = \left[0, \frac{1}{2}\right) \times [0, g)$ and

$$\mathcal{M}(\Delta^-) = \left[-\frac{1}{2}, -g^2\right) \times [0, g^2) \cup [-g^2, 0) \times [0, g^2].$$

See Figure 5.2.

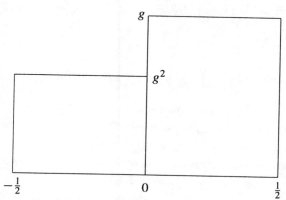

Figure 5.2.   The natural extension for the NICF

(b) Also determine Ω_S for $S = S_{\text{ocf}}$, for $S = S_g$ and more generally for $S = S_\alpha$, where $\frac{1}{2} \leq \alpha \leq 1$. See also [Nak81]. ∎

Remarks 5.4.25.

1. From Theorems 5.4.21 and 5.4.23 it follows that $(\Omega_S, \mathcal{L}, \mu_S, \mathcal{T}_S)$, which is the two-dimensional ergodic system underlying the corresponding S-expansion, is isomorphic (via the $\mathcal{M}$-map) to an induced system of (Ω, T) with return-time bounded by 2.

2. One can show that $\mathcal{T}_S$ can be written in the following way:

$$\mathcal{T}_S(t, v) = \left(\left| \frac{1}{t} \right| - f_S(t, v), \frac{1}{\mathrm{sgn}(t) \cdot v + f_S(t, v)} \right), \quad \text{for } (t, v) \in \Omega_S.$$

Furthermore one has

$$b_{k+1} = f_S(t_k, v_k), \ k \geq 0, \text{ where } (t_0, v_0) = (x - b_0, 0).$$

Thus we see that the S-expansion is the process associated with $\mathcal{T}_S$ and f_S. For details, see [Kra91].

For the aforementioned examples we have

$$f_{\frac{1}{2}}(t, v) = \left\lfloor \left| \frac{1}{t} \right| + \frac{1}{2} \right\rfloor \text{ (NICF), } \quad f_g(t, v) = \left\lfloor \left| \frac{1}{t} \right| + g^2 \right\rfloor \text{ (SCF)}$$

and

$$f_\alpha(t, v) = \left\lfloor \left| \frac{1}{t} \right| + 1 - \alpha \right\rfloor \quad (\alpha\text{-expansion) for } \frac{1}{2} \leq \alpha \leq 1.$$

3. In case of the OCF the last statement of Theorem 5.4.23 says that for a.e. $x \in [-\frac{1}{2}, \frac{1}{2})$ the sequence $\left(T_{\mathrm{ocf}}^n(x, 0) \right)_{n \geq 0}$ is distributed according to the density function $(\log G)^{-1}(1 + tv)^{-2}$; i.e., it behaves like the orbit of a generic point. In general, this last statement is a direct corollary of Jager's Lemma, Lemma 5.3.11. ∎

At this point one could mimic for any S-expansion what we did for the RCF in Sections 1.3 and 5.3. Matrices can be used to obtain recurrence relations for the numerators and denominators of S-convergents, the (distribution of) S-approximation coefficients can be studied, etc.

Although such a variation on a theme yields interesting results in itself, we refrain from doing it. We invite the interested reader to consult [Kra91] for further results and references.

5.5 A skew product related to continued fractions

In ergodic theory the idea of a *skew product* forms an important concept. In this final section on continued fractions we will not discuss skew products in general, but rather illustrate the concept by an example: a skew product related to continued fractions.

Let $m \in \mathbb{Z}$, $m \geq 2$, and let $G(m)$ be the group of 2×2 matrices of determinant ± 1 and with entries in the ring $\mathbb{Z}/m\mathbb{Z}$, i.e.,

$$G(m) := \left\{ \begin{pmatrix} \alpha & \beta \\ \gamma & \delta \end{pmatrix} ; \ \alpha, \beta, \gamma, \delta \in \mathbb{Z}/m\mathbb{Z}, \ |\alpha\delta - \beta\gamma| = 1 \right\}.$$

It is easily seen that $G(m)$ forms a finite group.

Exercise 5.5.1. Let $m \in \mathbb{Z}$, $m \geq 2$, and let $\mathrm{SL}_2(\mathbb{Z}/m\mathbb{Z})$ be given by

$$\mathrm{SL}_2(\mathbb{Z}/m\mathbb{Z}) := \{A \in G(m); \ \det(A) = 1\}.$$

(a) Show that $\mathrm{SL}_2(\mathbb{Z}/m\mathbb{Z})$ is a subgroup of $G(m)$ of index 2 if $m > 2$, and of index 1 if $m = 2$.

(b) From [Shi71]: Show that the cardinality of $\mathrm{SL}_2(\mathbb{Z}/m\mathbb{Z})$ equals $mJ(m)$. Here J is *Jordan's arithmetical totient function*, defined by

$$J(m) := m^2 \prod_{p|m} \left(1 - \frac{1}{p^2} \right),$$

where the product is taken over all the primes p that divide m. See also [Apo90], p. 48, and [PS64], VIII, Aufgabe 64.

(c) Conclude that the cardinality $|G(m)|$ of $G(m)$ equals

$$|G(m)| = \begin{cases} 2m\,J(m), & m > 2, \\ m\,J(m) = 6, & m = 2. \end{cases} \qquad \blacksquare$$

For the next theorem $G(m)$ needs to have a measure structure. A natural choice for the σ-algebra is the collection $\mathcal{D}(m)$ of all subsets of $G(m)$, and for the measure h_m the discrete uniform distribution (also known as *Haar measure*).

Theorem 5.5.2. (Jager and Liardet, 1988) *Let $m \in \mathbb{Z}$, $m \geq 2$, and let the transformation $\mathcal{J} : [0, 1) \times G(m) \to [0, 1) \times G(m)$ be defined by*

$$\mathcal{J}(x, g) := \left(Tx, \ g \begin{pmatrix} 0 & 1 \\ 1 & \overline{a_1(x)} \end{pmatrix} \right), \quad (x, g) \in [0, 1) \times G(m),$$

where T is the continued fraction map and the bar denotes reduction modulo m. Then the skew product

$$\Gamma := ([0, 1) \times G(m), \mathcal{L} \times \mathcal{D}(m), \mu \times h_m, \mathcal{J})$$

is ergodic.

Furthermore, if $(p_n/q_n)_{n \geq 1}$ is the sequence of RCF-convergents of x, then for almost all $x \in [0, 1)$ the sequence of matrices

$$n \mapsto \begin{pmatrix} \overline{p_{n-1}} & \overline{p_n} \\ \overline{q_{n-1}} & \overline{q_n} \end{pmatrix}, \quad n \geq 1,$$

is uniformly distributed over $G(m)$.

The proof of this theorem resembles to some extent the proof that the continued fraction map T is ergodic. Clearly the proof of the above theorem must be more complicated, since the dynamical system of the RCF is now skewed with $G(m)$.

One of the nice ingredients of the proof is the following lemma, which we mention just for its own beauty.

Lemma 5.5.3. *Let $m \in \mathbb{Z}$, $m \geq 2$, and let*

$$\begin{pmatrix} a & b \\ c & d \end{pmatrix}$$

be a matrix with integer entries and determinant ± 1. *Then there exists a finite sequence of positive integers* $a_1, \ldots, a_n$, *such that*

$$\begin{pmatrix} a & b \\ c & d \end{pmatrix} \equiv \begin{pmatrix} 0 & 1 \\ 1 & a_1 \end{pmatrix} \begin{pmatrix} 0 & 1 \\ 1 & a_2 \end{pmatrix} \cdots \begin{pmatrix} 0 & 1 \\ 1 & a_n \end{pmatrix} \pmod{m}.$$

For a proof of Theorem 5.5.2 we refer to [JL88], Section 4. From Theorem 5.5.2 and the Ergodic Theorem, Jager and Liardet were able to draw several corollaries, some of which were previously obtained (in a completely different way) by R. Moeckel [Moe82]. Here we mention the following result.

Proposition 5.5.4. (Moeckel 1982; Jager and Liardet, 1988) *Let* p, q *and* m *be three integers, such that* $m \geq 2$ *and* $(p, q, m) = 1$. *Then for almost all* x *one has*

$$\lim_{N \to \infty} \frac{1}{N} \# \left\{ n;\ 1 \leq n \leq N,\ \begin{pmatrix} p_n \\ q_n \end{pmatrix} \equiv \begin{pmatrix} p \\ q \end{pmatrix} \bmod m \right\} = \frac{1}{J(m)}.$$

Consider, modulo 2, the sequence $(q_n)_{n \geq -1}$ of the RCF-convergents of an irrational number x. Obviously this is a sequence of zeroes and ones. Again Moeckel [Moe82], and later Jager and Liardet [JL88], showed that for almost all x the blocks 01, 10 and 11 occur in this sequences with equal probabilities, i.e., $\frac{1}{3}$; notice that one can never have an occurrence of 00. Using the Jager-Liardet skew-product Γ, V. Nolte [Nol90] obtained (among other things) the probabilities of the five possible blocks of length three and of the eight possible blocks of length four in such sequences of zeroes and ones. Due to technical difficulties his method breaks down for longer blocks.

In 1998, the natural extension of Γ was studied in [DK98]. Due to this—and using a natural generalization of the concept of singularization—results similar to those described above could be transported to S-expansions. This leads to the surprising result that for any S-expansion and for almost all x the sequence of numerators $(r_n)_{n \geq 1}$ resp. denominators $(s_n)_{n \geq 1}$ of the S-convergents $(r_n/s_n)_{n \geq 1}$ of x have—mod m—the same asymptotic behavior as the sequence of numerators $(p_n)_{n\geq1}$ resp. denominators $(q_n)_{n\geq1}$ of the RCF-convergents $(p_n/q_n)_{n\geq1}$ of x.

5.6   For further reading

Section 5.4 deals with the theory one can build using singularizations. As we saw, a singularization is equivalent to removing a convergent. Can one add other rationals as convergents and still get a semi-regular continued fraction expansion? One can add *intermediate convergents* via a similar algebraic manipulation

$$ a + \cfrac{1}{b + \xi} = a + 1 + \cfrac{-1}{1 + \cfrac{1}{b - 1 + \xi}}. $$

By going in the other direction, one can build a theory of *insertions*, similar to the theory of singularizations. In the same way as singularizations are related to the induced transformations of subsection 4.2.1, insertions are related to the integral transformations of subsection 4.2.2. We decided not to go into these details; continued fractions would take over this book! However, the interested reader is referred to [DK00].

CHAPTER 6
Entropy

6.1   Introduction

6.1.1   Randomness and information

Given a measure preserving transformation T on a probability space $(X, \mathcal{F}, \mu)$, we want to define a nonnegative quantity $h(T)$ that measures the average uncertainty about where T moves the points of X. That is, the value of $h(T)$ reflects the amount of randomness generated by T. We want to define $h(T)$ in such a way that (i) the amount of information gained by an application of T is proportional to the amount of uncertainty removed, and (ii) $h(T)$ is isomorphism invariant, so that isomorphic transformations have equal entropy.

The connection between entropy (that is randomness, uncertainty) and the transmission of information was first studied by Claude Shannon in 1948. As a motivation let us look at the following simple example. Consider a source (for example a ticker-tape) that produces a string of symbols $\ldots x_{-1} x_0 x_1 \ldots$ from the alphabet $\{a_1, a_2, \ldots, a_n\}$. Suppose that the probability of receiving symbol a_i at any given time is p_i, and that each symbol is transmitted independently of what has been transmitted earlier. Of course we must have here that each $p_i \geq 0$ and that $\sum_i p_i = 1$. In ergodic theory we view this process as the dynamical system $(X, \mathcal{F}, \mu, T)$, where $X = \{a_1, a_2, \ldots, a_n\}^{\mathbb{Z}}$, $\mathcal{F}$ is the σ-algebra generated by cylinder sets of the form

$$\Delta_n(a_{i_1}, a_{i_2}, \ldots, a_{i_n}) := \{x \in X : x_{i_1} = a_{i_1}, \ldots, x_{i_n} = a_{i_n}\},$$

μ is the product measure assigning to each coordinate the probability p_i of seeing the symbol a_i, and T is the left shift. In case our sequence of symbols is one-sided we take $X = \{a_1, a_2, \ldots, a_n\}^{\mathbb{N}}$. We define the entropy of this system by

$$H(p_1, \ldots, p_n) := -\sum_{i=1}^{n} p_i \log p_i . \qquad (6.1)$$

If we define $\log p_i$ as the amount of uncertainty in transmitting the symbol a_i, then H is the average amount of information (or uncertainty) per symbol (notice that H is in fact an expected value). To see why this is an appropriate definition, notice that if the source is degenerate, that is, $p_i = 1$ for some i (i.e., the source transmits only the symbol a_i), then $H = 0$. In this case we indeed have no randomness. Another reason this definition is appropriate is that H is maximal if $p_i = \frac{1}{n}$ for all i, and this agrees with the fact that the source is most random when all the symbols are equiprobable. To see this maximum, consider the function $f : [0, 1] \to \mathbb{R}_+$ defined by

$$f(t) := \begin{cases} 0, & \text{if } t = 0, \\ -t \log t, & \text{if } 0 < t \leq 1. \end{cases}$$

Then f is continuous and concave downward, and Jensen's Inequality implies that for any $p_1, \ldots, p_n$ with $p_i \geq 0$ and $p_1 + \cdots + p_n = 1$,

$$\frac{1}{n} H(p_1, \ldots, p_n) = \frac{1}{n} \sum_{i=1}^{n} f(p_i) \leq f\left(\frac{1}{n} \sum_{i=1}^{n} p_i\right)$$

$$= f\left(\frac{1}{n}\right) = \frac{1}{n} \log n ,$$

so $H(p_1, \ldots, p_n) \leq \log n$ for all probability vectors $(p_1, \ldots, p_n)$. Since

$$H\left(\frac{1}{n},\ldots,\frac{1}{n}\right) = \log n,$$

the maximum value is attained at $(\frac{1}{n},\ldots,\frac{1}{n})$.

6.1.2 Definitions

So far H is defined as the average information per symbol. The above definition can be extended to define the information transmitted by the occurrence of an event E as $-\log\mu(E)$. This definition has the property that the information transmitted by $E \cap F$ for independent events E and F is the sum of the information transmitted by each one individually, i.e.,

$$-\log\mu(E \cap F) = -\log\mu(E) - \log\mu(F).$$

The only measurable function with this property is $\log_a x$, with $a > 0$. Since entropy originates from Information Theory (information is transmitted via binary sequences), some authors prefer to use $a = 2$. We prefer to use natural logarithms because it is more appropriate for our calculations.

In the above example of the ticker-tape, the symbols were transmitted independently. In general, the symbol generated might depend on what has been received before. In fact, these dependencies are often built-in to be able to check the transmitted sequence of symbols for errors (think here of the Morse sequence, sequences on compact discs, etc.). Such dependencies must be taken into consideration in the calculation of the average information per symbol. This can be achieved if one replaces the symbols a_i by blocks of symbols of a particular size. More precisely, for every n, let C_n be the collection of all possible n-blocks (or cylinder sets) of length n, and define

$$H_n = -\sum_{C \in C_n} P(C) \log P(C).$$

Then $\frac{1}{n} H_n$ can be viewed as the average information per symbol when a block of length n is transmitted. The entropy of the source is now defined by

$$h = \lim_{n \to \infty} \frac{H_n}{n} . \qquad (6.2)$$

The limit in (6.2) exists because H_n is a *subadditive sequence*, i.e., $H_{n+m} \le H_n + H_m$, and because of the following proposition.

Proposition 6.1.1. *If $\{a_n\}$ is a subadditive sequence of nonnegative real numbers, then*

$$\lim_{n \to \infty} \frac{a_n}{n}$$

exists.

Proof. Fix any $m > 0$. For any $n \ge 1$ one has $n = km + i$ for some i between $0 \le i \le m - 1$. By subadditivity it follows that

$$\frac{a_n}{n} = \frac{a_{km+i}}{km+i} \le \frac{a_{km}}{km} + \frac{a_i}{km} \le k\frac{a_m}{km} + \frac{a_i}{km} = \frac{a_m}{m} + \frac{a_i}{km} .$$

Note that if $n \to \infty$, $k \to \infty$ and so $\limsup_{n \to \infty} a_n / n \le a_m / m$. Since m is arbitrary one has

$$\limsup_{n \to \infty} \frac{a_n}{n} \le \inf_{n \ge 1} \frac{a_n}{n} \le \liminf_{n \to \infty} \frac{a_n}{n} .$$

Therefore $\lim_{n \to \infty} a_n / n$ exists, and equals $\inf a_n / n$. ∎

Now replace the source by a measure preserving system $(X, \mathcal{F}, \mu, T)$. How can one define the entropy of this system similar to the case of a source? The symbols $\{a_1, \dots, a_n\}$ can now be viewed as a partition $\alpha = \{A_1, \dots, A_n\}$ of X, so that X is the disjoint union (up to sets of measure zero) of $A_1, \dots, A_n$. The source can be seen as follows: with each point $x \in X$, we associate an infinite sequence $\dots, x_{-1}, x_0, x_1, \dots$, where x_i is a_j if and only if $T^i x \in A_j$. We define the *entropy of the partition* α by

$$H(\alpha) = H_\mu(\alpha) := -\sum_{i=1}^{n} \mu(A_i) \log \mu(A_i) .$$

Our aim is to define the entropy $h(T)$ of the (not necessarily invertible) transformation T independent of the partition we choose. In fact $h(T)$ will be the supremum of the entropies over all possible finite partitions. But first we need a few facts about partitions.

Exercise 6.1.2. Let $\alpha = \{A_1, \ldots, A_n\}$ and $\beta = \{B_1, \ldots, B_m\}$ be two partitions of X. Show that

$$T^{-1}\alpha := \{T^{-1}A_1, \ldots, T^{-1}A_n\}$$

and

$$\alpha \vee \beta := \{A_i \cap B_j : A_i \in \alpha,\ B_j \in \beta\}$$

are both partitions of X.   ∎

The members of a partition are called the *atoms* of the partition. We say that the partition β is a *refinement* of the partition α, and write $\alpha \leq \beta$, if for every $1 \leq j \leq m$ there exists $1 \leq i \leq n$ such that $B_j \subset A_i$ (up to sets of measure zero). The partition $\alpha \vee \beta$ is called the *common refinement* of α and β.

Exercise 6.1.3. Show that if β is a refinement of α, then each atom of α is a finite (disjoint) union of atoms of β.   ∎

Exercise 6.1.4. Let α and β be partitions of $(X, \mathcal{F}, \mu, T)$, where T is a measure preserving transformation.

(a) Show that $H(T^{-1}\alpha) = H(\alpha)$.

(b) Show that if $\alpha \leq \beta$, then $H(\alpha) \leq H(\beta)$.

(c) Show that $H(\alpha \vee \beta) \leq H(\alpha) + H(\beta)$.

(d) We call two partitions α and β *independent* if

$$\mu(A \cap B) = \mu(A)\mu(B) \text{ for all } A \in \alpha,\ B \in \beta.$$

Show that, if α and β are independent partitions,

$$H(\alpha \vee \beta) = H(\alpha) + H(\beta) . \qquad \blacksquare$$

Now consider the partition $\bigvee_{i=0}^{n-1} T^{-i}\alpha$, whose atoms are sets of the form $A_{i_0} \cap T^{-1}A_{i_1} \cap \cdots \cap T^{-(n-1)}A_{i_{n-1}}$, consisting of all points $x \in X$ with the property that $x \in A_{i_0}$, $Tx \in A_{i_1}, \ldots, T^{n-1}x \in A_{i_{n-1}}$. We are now in position to give the definition of the entropy of the transformation T.

Definition 6.1.5. *The entropy of the measure preserving transformation T with respect to the partition α is given by*

$$h(\alpha, T) = h_\mu(\alpha, T) := \lim_{n \to \infty} \frac{1}{n} H\left(\bigvee_{i=0}^{n-1} T^{-i}\alpha\right), \qquad (6.3)$$

where

$$H\left(\bigvee_{i=0}^{n-1} T^{-i}\alpha\right) = - \sum_{D \in \bigvee_{i=0}^{n-1} T^{-i}\alpha} \mu(D) \log(\mu(D)) .$$

Finally, the entropy of the transformation T is given by

$$h(T) = h_\mu(T) := \sup_\alpha h(\alpha, T) ,$$

where the supremum is taken over all finite partitions α of finite entropy.

Remarks 6.1.6.

1. Note that the limit in (6.3) exists due to Proposition 6.1.1.

2. In almost all of the classical books on ergodic theory, entropy is defined as above, i.e., via finite partitions. For a number of the transformations we are interested in, like the GLS transformations and the continued fraction map, it is more natural to work with countable partitions. Fortunately, in the above definition of entropy, and in the the-

orems that will follow, one can replace finite partitions by countable partitions of finite entropy. See [Bro76], p. 148, and [Pet89], p. 248.

∎

We have the following theorem.

Theorem 6.1.7. *Entropy is an isomorphism invariant.*

Proof. Let $(X, \mathcal{F}, \mu, T)$ and $(Y, \mathcal{C}, \nu, S)$ be two isomorphic measure preserving systems (see Definition 1.2.18), with $\psi : X \to Y$ the corresponding isomorphism. We need to show that $h_\mu(T) = h_\nu(S)$.

Let $\beta = \{B_1, \ldots, B_n\}$ be any partition of Y; then $\psi^{-1}\beta = \{\psi^{-1}B_1, \ldots, \psi^{-1}B_n\}$ is a partition of X. Set $A_i = \psi^{-1}B_i$, for $1 \leq i \leq n$. Since $\psi : X \to Y$ is an isomorphism, we have that $\nu = \mu\psi^{-1}$ and $\psi T = S\psi$, so that for any $n \geq 0$ and $B_{i_0}, \ldots, B_{i_{n-1}} \in \beta$

$$\nu \left(B_{i_0} \cap S^{-1}B_{i_1} \cap \cdots \cap S^{-(n-1)}B_{i_{n-1}} \right)$$

$$= \mu \left(\psi^{-1}B_{i_0} \cap \psi^{-1}S^{-1}B_{i_1} \cap \cdots \cap \psi^{-1}S^{-(n-1)}B_{i_{n-1}} \right)$$

$$= \mu \left(\psi^{-1}B_{i_0} \cap T^{-1}\psi^{-1}B_{i_1} \cap \cdots \cap T^{-(n-1)}\psi^{-1}B_{i_{n-1}} \right)$$

$$= \mu \left(A_{i_0} \cap T^{-1}A_{i_1} \cap \cdots \cap T^{-(n-1)}A_{i_{n-1}} \right).$$

Setting

$$A(n) = A_{i_0} \cap \cdots \cap T^{-(n-1)}A_{i_{n-1}}$$

and

$$B(n) = B_{i_0} \cap \cdots \cap S^{-(n-1)}B_{i_{n-1}},$$

we thus find that

$$h_\nu(S) = \sup_\beta h_\nu(\beta, S) = \sup_\beta \lim_{n \to \infty} \frac{1}{n} H \left(\bigvee_{i=0}^{n-1} S^{-i}\beta \right)$$

$$= \sup_{\beta} \lim_{n \to \infty} -\frac{1}{n} \sum_{B(n) \in \bigvee_{i=0}^{n-1} S^{-i}\beta} \nu(B(n)) \log \nu(B(n))$$

$$= \sup_{\psi^{-1}\beta} \lim_{n \to \infty} -\frac{1}{n} \sum_{A(n) \in \bigvee_{i=0}^{n-1} T^{-i}\psi^{-1}\beta} \mu(A(n)) \log \mu(A(n))$$

$$= \sup_{\psi^{-1}\beta} h_{\mu}(\psi^{-1}\beta, T)$$

$$\leq \sup_{\alpha} h_{\mu}(\alpha, T) = h_{\mu}(T),$$

where in the last inequality the supremum is taken over all possible finite partitions α of X. Thus $h_{\nu}(S) \leq h_{\mu}(T)$. By symmetry $h_{\mu}(T) \leq h_{\nu}(S)$. Therefore $h_{\nu}(S) = h_{\mu}(T)$, and the proof is complete. ∎

Remark 6.1.8. Entropy is also preserved under the operation of taking natural extension; i.e., a measure preserving transformation has the same entropy as its natural extension. See [Bro76], p. 125. ∎

6.1.3   Calculation of entropy

Calculating the entropy of a transformation directly from the definition does not seem feasible, for one needs to take the supremum over *all* finite partitions, which is practically impossible. However, the entropy of a partition is relatively easier to calculate if one has full information about the partition under consideration. So the question arises whether it is possible to find a partition α of X where $h(\alpha, T) = h(T)$. Naturally, such a partition would contain all the information transmitted by T. To answer this question we need some notations and definitions. For simplicity we assume in this subsection that T is invertible.

For $\alpha = \{A_1, \dots, A_\ell\}$ and all $m > n \geq 0$, let

$$\alpha_n^m = \bigvee_{k=n}^{m} T^{-i}\alpha \quad \text{and} \quad \alpha_{-m}^{-n} = \bigvee_{k=-m}^{-n} T^{-i}\alpha,$$

and let $\sigma\left(\bigvee_{i=-\infty}^{\infty} T^{-i}\alpha\right)$ be the smallest σ-algebra containing all the partitions α_n^m and α_{-m}^{-n} for all n and m. In general we call a finite or countable partition α a *generator with respect to* T if $\sigma\left(\bigvee_{i=-\infty}^{\infty} T^{-i}\alpha\right) = \mathcal{F}$, where $\mathcal{F}$ is the σ-algebra on X. Naturally, this equality is modulo sets of measure zero. One also has the following characterization of generators, saying basically that each measurable set in X can be approximated by a finite disjoint union of cylinder sets. See also [Wal82], p. 6, for more details and proofs.

Proposition 6.1.9. *If the partition α is a generator of $\mathcal{F}$, then for each $A \in \mathcal{F}$ and for each $\varepsilon > 0$ there exist $m > n \geq 0$ and a finite disjoint union C of elements of $\{\alpha_n^m\}$, such that $\mu(A\triangle C) < \varepsilon$.*

We now state two famous theorems known as *Kolmogorov-Sinai's Theorem* and *Krieger's Generator Theorem*. We invite the interested reader to refer to Petersen's [Pet89] or Walter's [Wal82] books for the proofs.

Theorem 6.1.10. (Kolmogorov and Sinai, 1958) *If α is a finite or countable generator for T with $H(\alpha) < \infty$, then $h(T) = h(\alpha, T)$.*

The theorem of Kolmogorov and Sinai gives a convenient way of calculating the entropy of a transformation.

Theorem 6.1.11. (Krieger, 1970) *If T is an ergodic measure preserving transformation with $h(T) < \infty$, then T has a finite generator.*

We will use these two theorems to calculate the entropy of a Bernoulli shift, which agrees with formula (6.1).

Example 6.1.12. Consider the two-sided Bernoulli shift on the symbols $1, 2, \ldots, n$ (with $n \in \mathbb{N} \cup \{\infty\}$), and with weights given by the probability vector $(p_1, p_2, \ldots, p_n)$, satisfying $\sum_i -p_i \log p_i < \infty$.

In other words, we have the measure preserving system $(X, \mathcal{F}, \mu, T)$, where $X = \{1, 2, \ldots, n\}^{\mathbb{Z}}$ and $\mathcal{F}$ is the σ-algebra generated by the cylinder sets of the form

$$\{x \in X : x_0 = i_0, \ldots, x_m = i_m\}, \ i_j \in \{1, 2, \ldots, n\}.$$

This means that every measurable set in X can be approximated by a finite disjoint union of cylinders; see also Proposition 6.1.9. The measure μ is the product measure given by

$$\mu(\{x \in X : x_0 = i_0, \ldots, x_m = i_m\}) = p_{i_0} p_{i_1} \cdots p_{i_m},$$

and T as usual is the left shift. Our aim is to calculate $h(T)$. To this end we need to find a partition α that generates the σ-algebra $\mathcal{F}$ under the action of T. The natural choice of α is what is known as the *time-zero partition* $\alpha = \{A_1, \ldots, A_n\}$, where

$$A_i := \{x \in X : x_0 = i\}, \ i = 1, \ldots, n.$$

Notice that for all $m \in \mathbb{Z}$,

$$T^{-m} A_i = \{x \in X : x_m = i\},$$

and

$$A_{i_0} \cap T^{-1} A_{i_1} \cap \cdots \cap T^{-m} A_{i_m} = \{x \in X : x_0 = i_0, \ldots, x_m = i_m\}.$$

In other words, $\bigvee_{i=0}^{m} T^{-i} \alpha$ is precisely the collection of cylinders of length m (i.e., the collection of all m-blocks), and by definition these generate $\mathcal{F}$. Hence, α is a generating partition, so that

$$h(T) = h(\alpha, T) = \lim_{m \to \infty} \frac{1}{m} H\left(\bigvee_{i=0}^{m-1} T^{-i} \alpha\right).$$

First notice that—since μ is product measure here—the partitions

$$\alpha, \ T^{-1}\alpha, \ldots, \ T^{-(m-1)}\alpha$$

are all independent, since each specifies a different coordinate, and so

$$H(\alpha \vee T^{-1}\alpha \vee \cdots \vee T^{-(m-1)}\alpha)$$

$$= H(\alpha) + H(T^{-1}\alpha) + \cdots + H(T^{-(m-1)}\alpha)$$

$$= m H(\alpha) = -m \sum_{i=1}^{n} p_i \log p_i.$$

For the last step, see also Exercise 6.1.4(a,d). Thus,

$$h(T) = \lim_{m \to \infty} \frac{1}{m}(-m) \sum_{i=1}^{n} p_i \log p_i = -\sum_{i=1}^{n} p_i \log p_i. \qquad \blacksquare$$

Corollary 6.1.13. *If T is a GLS($\mathcal{I}$) transformation with $\mathcal{I} = \{I_n : n \in \mathcal{D}\}$, where $\mathcal{D} \subset \mathbb{N}$ is the digit set, $\lambda(I_n) := L_n$, $\sum_{n \in \mathcal{D}} L_n = 1$, and $H(\mathcal{I}) < \infty$, then $h(T) = -\sum_{n \in \mathcal{D}} L_n \log L_n$.*

Proof. We have seen earlier (see also Section 4.4) that if we identify points with infinite sequences of symbols from the digit set $\mathcal{D}$, then the natural extension map $\mathcal{T}$ of T can be seen as a left shift, and

$$\lambda(\{x : x_0 = i_0, \ldots, x_m = i_m\}) = L_{i_0} \cdots L_{i_m}.$$

Further, λ can be viewed as product measure, and so the dynamical system $([0, 1), \mathcal{B}, \lambda, \mathcal{T})$ is a Bernoulli shift with weights $\{L_n : n \in \mathcal{D}\}$; hence we have $h(\mathcal{T}) = -\sum_{n \in \mathcal{D}} L_n \log L_n$. $\qquad \blacksquare$

6.1.4 Conditional entropy

Given two partitions $\alpha = \{A_1, \ldots, A_n\}$ and $\beta = \{B_1, \ldots, B_m\}$ of X, and under the convention that $0 \log 0 := 0$, we define the *conditional entropy of α given β* by

$$H(\alpha|\beta) := -\sum_{A \in \alpha} \sum_{B \in \beta} \log\left(\frac{\mu(A \cap B)}{\mu(B)}\right) \mu(A \cap B).$$

Exercise 6.1.14. Let $\alpha = \{A_1, \ldots, A_n\}$ and $\beta = \{B_1, \ldots, B_m\}$ be two independent partitions of X. Then show that $H(\alpha|\beta) = H(\alpha)$.

$\blacksquare$

The above quantity $H(\alpha|\beta)$ is interpreted as the average uncertainty about which element of the partition α the point x will enter (under T) if we already know which element of β the point x will enter.

Exercise 6.1.15. Let α, β and γ be partitions of X.

(a) Show that $H(\alpha \vee \beta|\gamma) = H(\alpha|\gamma) + H(\beta|\alpha \vee \gamma)$.

(b) Show that, if $\beta \leq \alpha$, then $H(\gamma|\alpha) \leq H(\gamma|\beta)$.

(c) Show that $H(\alpha \vee \beta) = H(\alpha) + H(\beta|\alpha)$.

(d) Show that, if $\beta \leq \alpha$, then $H(\beta|\alpha) = 0$.

(e) Explain in words why each result (a)–(d) is reasonable.   ∎

Using induction and Exercise 6.1.15 part (c) repeatedly, one obtains the following theorem.

Theorem 6.1.16. *The entropy of the measure preserving transformation T with respect to the partition α is also given by*

$$h(\alpha, T) = \lim_{n \to \infty} H\left(\alpha \mid \bigvee_{i=1}^{n-1} T^{-i}\alpha\right).$$

6.1.5   Entropy of β-transformations

In section 4.5, we showed that any β-transformation contains a Bernoulli shift as an induced system, namely an appropriate $\mathrm{GLS}(\mathcal{I})$ transformation. Example 6.1.12 showed how one can calculate, in a straightforward way, the entropy of Bernoulli shifts. The following theorem, known as *Abramov's formula*, gives the relationship between the entropy of an induced transformation and the original map. For a proof, the reader is referred to [Pet89].

Theorem 6.1.17. (Abramov's formula) *Let $(X, \mathcal{F}, \mu, T)$ be a measure preserving dynamical system. Let $B \in \mathcal{F}$ be of positive measure, and* *denote by $T_B : B \to B$ the corresponding induced transformation. Then*

$$h(T_B) = \frac{h(T)}{\mu(B)}. \tag{6.4}$$

We calculate the entropy of a β-transformation using Abramov's formula. For simplicity we start with the case that β is a pseudo-golden mean number with $m = 3$. We use the notations given in Section 4.5.1. We have seen that in this case the natural extension of T_β is given by $(X, \mathcal{E}, \mu, \mathcal{T}_\beta)$, where

$$X = \bigcup_{k=0}^{2} \left[T_\beta^{3-k} 1, \ T_\beta^{3-k-1} 1 \right) \times \left[0, T_\beta^k 1 \right),$$

$\mathcal{E}$ is the restriction of the product Lebesgue σ-algebra $\mathcal{L} \times \mathcal{L}$ on X,

$$\mathcal{T}_\beta(x, y) := \left(T_\beta x, \frac{1}{\beta}(\lfloor \beta x \rfloor + y) \right) \quad \text{for } (x, y) \in X,$$

and

$$\mu(A) = \frac{\beta}{\dfrac{1}{\beta} + \dfrac{2}{\beta^2} + \dfrac{3}{\beta^3}} (\lambda \times \lambda)(A)$$

for any measurable subset A of X. Further, the induced system $(Y, \mathcal{W}, \rho)$ with $Y = [0, 1) \times [0, 1/\beta]$, $\mathcal{W} = (\mathcal{T}_\beta)_Y$ and $\rho = \beta(\lambda \times \lambda)$ was shown to be isomorphic to the natural extension $\mathcal{S}$ of the GLS-transformation S given by

$$Sx = \begin{cases} \beta x, & x \in [0, \frac{1}{\beta}), \\ \beta^2 x - \beta, & x \in [\frac{1}{\beta}, \frac{1}{\beta} + \frac{1}{\beta^2}), \\ \beta^3 x - \beta^2 - \beta, & x \in [\frac{1}{\beta} + \frac{1}{\beta^2}, 1). \end{cases}$$

Since every GLS-transformation is a Bernoulli shift, it follows that $\mathcal{S}$ is isomorphic to the two-sided Bernoulli shift with weights $(\frac{1}{\beta}, \frac{1}{\beta^2}, \frac{1}{\beta^3})$. Hence by Example 6.1.12, the entropy of $\mathcal{S}$—and therefore also of $\mathcal{W}$ (due to Theorem 6.1.7)—equals

$$\left(\frac{1}{\beta} + \frac{2}{\beta^2} + \frac{3}{\beta^3} \right) \log \beta \, .$$

Now

$$\mu(Y) = \frac{\beta}{\dfrac{1}{\beta} + \dfrac{2}{\beta^2} + \dfrac{3}{\beta^3}} \int_0^1 \int_0^{\frac{1}{\beta}} dx \, dy = \frac{1}{\dfrac{1}{\beta} + \dfrac{2}{\beta^2} + \dfrac{3}{\beta^3}} \, ,$$

and so by Abramov's formula (6.4)

$$h(T_\beta) = \frac{1}{\mu(Y)} h(\mathcal{W}) = \log \beta \, .$$

Due to Remark 6.1.8 (i.e., entropy is preserved under the operation of taking a natural extension), we also have that $h(T_\beta) = \log \beta$.

Performing similar calculations yields that $h(T_\beta) = \log \beta$ for any pseudo-golden mean number β.

In fact, more is true and can be proved in a similar way.

Theorem 6.1.18. *For any $\beta > 1$, $h(T_\beta) = \log \beta$.*

Proof. For the proof we use the same notation as in Section 4.5.2. Since

$$1 = \sum_{k=1}^{\infty} \frac{b_k}{\beta^k} \quad \text{and} \quad T_\beta^i 1 = \sum_{k=1}^{\infty} \frac{b_{k+i}}{\beta^k} \, ,$$

it follows that

$$\eta(\mathcal{H}_\beta) = 1 + \sum_{i=1}^{\infty} \frac{T_\beta^i 1}{\beta^i} = 1 + \sum_{i=1}^{\infty} \sum_{k=1}^{\infty} \frac{b_{k+i}}{\beta^{k+i}} \, .$$

Since $b_n \leq \lfloor \beta \rfloor$ for all $n \geq 1$, it follows that $\displaystyle\sum_{i=1}^{\infty} \sum_{k=1}^{\infty} \frac{b_{k+i}}{\beta^{k+i}}$ is convergent, and that

$$\sum_{i=1}^{\infty} \sum_{k=1}^{\infty} \frac{b_{k+i}}{\beta^{k+i}} = \frac{b_2}{\beta^2} + \frac{b_3}{\beta^3} + \frac{b_4}{\beta^4} + \cdots$$

$$+ \frac{b_3}{\beta^3} + \frac{b_4}{\beta^4} + \frac{b_5}{\beta^5} + \cdots$$

$$+ \frac{b_4}{\beta^4} + \frac{b_5}{\beta^5} + \frac{b_6}{\beta^6} + \cdots$$

$$\vdots$$

Summing over the anti-diagonals now yields that

$$\eta(\mathcal{H}_\beta) = 1 + \sum_{k=1}^{\infty} k \frac{b_{k+1}}{\beta^{k+1}} = \sum_{k=0}^{\infty} (k+1) \frac{b_{k+1}}{\beta^{k+1}} .$$

Now we consider the induced system $([0, 1]^2, \mathcal{W}_\beta)$ of $\mathcal{T}_\beta$, which is isomorphic to the natural extension of the GLS-transformation with partition as given in Equation (4.5). Thus

$$h(\mathcal{W}_\beta) = - \sum_{k=0}^{\infty} \frac{b_{k+1}}{\beta^{k+1}} \log \frac{1}{\beta^{k+1}}$$

$$= \log \beta \sum_{k=0}^{\infty} (k+1) \frac{b_{k+1}}{\beta^{k+1}} .$$

From this and Abramov's formula it follows that $h(\mathcal{T}_\beta) = \log \beta$. Again due to Remark 6.1.8, we have that $h(T_\beta) = \log \beta$. ∎

6.2 The Shannon-McMillan-Breiman Theorem and some consequences

6.2.1 The Shannon-McMillan-Breiman Theorem

Let us consider an ergodic measure preserving system $(X, \mathcal{F}, \mu, T)$, and let α be a finite or countable partition of X. We have seen that for each n, $\bigvee_{i=0}^{n-1} T^{-i} \alpha$ is again a partition of X, and hence a.e. x belongs to a unique *atom* $A_n \in \bigvee_{i=0}^{n-1} T^{-i}\alpha$. In this case, i.e., in case $x \in A_n$, we denote A_n by $A_n(x)$. Of course, $A_n(x)$ is what we call a cylinder set (with respect to partition α) of size n containing x. The Shannon-McMillan-Breiman Theorem gives the asymptotic size of such cylinders.

Theorem 6.2.1. (Shannon-McMillan-Breiman) *Let* $(X, \mathcal{F}, \mu, T)$ *be an ergodic measure preserving system, and let* α *be a finite or countable partition of X with* $H(\alpha) < \infty$. *Then for a.e.* $x \in X$,

$$\lim_{n \to \infty} -\frac{1}{n} \log \left(\mu(A_n(x)) \right) = h(\alpha, T).$$

The interested reader can find a proof of this theorem in any standard book on ergodic theory, e.g., [Bil65], [CFS82], [Pet89], [Wal82].

The entropy $h(T)$ of the regular continued fraction map T is easily obtained from Lévy's Proposition 3.5.5, the Theorem of Shannon-McMillan-Breiman and the following lemma.

Lemma 6.2.2. *Let* $\Delta_n(x)$ *denote the cylinder set of order n for the continued fraction map T containing x. Then for all irrational x*

$$\lim_{n \to \infty} \frac{\log \lambda(\Delta_n(x))}{\log \mu(\Delta_n(x))} = 1.$$

Proof. From Equation (3.4),

$$\frac{1}{2 \log 2} \lambda(\Delta_n(x)) \le \mu(\Delta_n(x)) \le \frac{1}{\log 2} \lambda(\Delta_n(x)).$$

It follows that

$$\frac{\log(2 \log 2) + \log \mu(\Delta_n(x))}{\log \mu(\Delta_n(x))} \le \frac{\log \lambda(\Delta_n(x))}{\log \mu(\Delta_n(x))}$$

$$\le \frac{\log(\log 2) + \log \mu(\Delta_n(x))}{\log \mu(\Delta_n(x))}.$$

Taking limits, the desired result follows from the fact that $\mu(\Delta_n(x)) \to 0$ as $n \to \infty$. ∎

Remark 6.2.3. Here we specified the lemma for the continued fraction map, but it holds in general for any measure preserving transformation with a measure equivalent to Lebesgue measure λ, which has a density which is bounded away from 0 and ∞. In the general case one has to replace $\Delta_n(x)$ in the statement of the lemma by $(A_{i_0} \cap T^{-1}A_{i_1} \cap \cdots \cap T^{-(n-1)}A_{i_{n-1}})(x)$, the atom containing x, where $A_1, \ldots, A_m$ forms a generating partition. ∎

It follows from Lemma 6.2.2 that for almost all x

$$\lim_{n\to\infty} -\frac{1}{n} \log \lambda(\Delta_n(x)) = \lim_{n\to\infty} -\frac{1}{n} \log \mu(\Delta_n(x)).$$

In view of the Theorem of Shannon-McMillan-Breiman it thus suffices to determine the former limit for almost all x.

Exercise 6.2.4. Using Equation (3.8) in Proposition 3.5.5 and Lemma 6.2.2, show that

$$h(T) = \frac{\pi^2}{6\log 2}.$$ ∎

Exercise 6.2.5. Let $S \subset [0, 1) \times [0, 1]$ be a singularization area; see also Section 5.4. Use Abramov's formula (6.4) to show that

$$h(T_S) = \frac{1}{1 - \mu(S)} \frac{\pi^2}{6\log 2}.$$ ∎

Remark 6.2.6. Without going into details, we want to mention here that there exists a class of maps T (containing all the examples from this book like n-ary expansions, continued fractions and β-expansions), for which the entropy $h(T)$ can easily be calculated using the *Rohlin Entropy Formula*,

$$h(T) = \int \log |T'(x)| d\mu(x);$$

here T' denotes derivative. Essential ingredients of the proof of Rohlin's Entropy Formula are Lemma 6.2.2, the Theorem of Shannon-McMillan-Breiman, and the so-called *Rényi Condition*, which follows if one wants to generalize the proof of the ergodicity of the (regular) continued fraction map to more general maps; see also [Rén57]. For a proof of Rohlin's Entropy Formula we refer to the book of M. Pollicott and M. Yuri [PY98], p. 133.   ∎

Exercise 6.2.7. Use Rohlin's Entropy Formula to calculate the entropy $h(T)$, when T is the map corresponding to

(a) n-ary expansions, for $n \in \mathbb{N}, n \geq 2$;

(b) β expansions, for $\beta > 1, \beta \notin \mathbb{N}$;

(c) the regular continued fraction.   ∎

6.2.2 Lochs' Theorem

Suppose that the irrational number x has decimal expansion $x = .d_1 d_2 \ldots$, and RCF-expansion (1.6). Let y be the rational number determined by the first n digits of x, i.e., $y = .d_1 d_2 \ldots d_n$, and let $[0; c_1, c_2, \ldots, c_k]$ be the RCF-expansion of y. We now define the function $m = m(n, x)$ as follows: m is the largest positive integer for which $a_i = c_i$ for $i \leq m$, i.e.,

$$a_1 = c_1, a_2 = c_2, \ldots, a_m = c_m, a_{m+1} \neq c_{m+1}.$$

In spite of results like Legendre's Theorem, Corollary 5.1.8, G. Lochs [Loc64] obtained the following, surprising result.

Theorem 6.2.8. (Lochs, 1962) *For almost all x one has*

$$\lim_{n \to \infty} \frac{m(n, x)}{n} = \frac{6 \log 2 \log 10}{\pi^2} = 0.97027014 \ldots.$$

Remark 6.2.9. Lochs' original proof rests on P. Lévy's Proposition 3.5.5. In this subsection we will see that Lochs' result is related to the ratio of entropies of the maps under consideration and is really a consequence of the Theorem of Shannon-McMillan-Breiman.   ∎

The continued fraction map has three essential properties that allow us to give a new proof of Lochs' result with the help of the Theorem of Shannon-McMillan-Breiman, Theorem 6.2.1. With this new approach we will be able to generalize Lochs' Theorem to all piecewise defined transformations with the same essential properties as the continued fraction.

(i) Firstly, the Gauss measure μ as defined in Section 1.3.3 has a bounded density

$$\frac{1}{\log 2}\frac{1}{1+x}$$

on $[0, 1)$, which implies equation (3.4). Due to this, Lemma 6.2.2 follows.

(ii) Secondly, for any cylinder set $\Delta_n = \Delta(a_1, \dots, a_n)$ one has that

$$\lambda(\Delta_n) = \frac{1}{q_n(q_n + q_{n-1})}.$$

(See also Exercise 1.3.15.) This, together with the recurrence relations in Exercise 1.3.8, yields the following result.

Exercise 6.2.10. Define $\Delta_n^+ := \Delta(a_1, \dots, a_{n-1}, a_n + 1)$. We call Δ_n and Δ_n^+ *adjacent cylinders*. Show that

$$\lambda(\Delta_n) \leq 3\lambda(\Delta_n^+).$$ ∎

Remark 6.2.11. In case $a_n \geq 2$, we also call $\Delta(a_1, \dots, a_{n-1}, a_n - 1)$ an adjacent cylinder to Δ_n, denoted by Δ_n^-.

(iii) Thirdly, the continued fraction map $Tx = \frac{1}{x} - \lfloor\frac{1}{x}\rfloor$ is decreasing on each partition element $\Delta_1(k)$, $k \in \mathbb{N}$. As a result, to obtain $\Delta_{n+1} = \Delta(a_1, \dots, a_{n+1})$ from $\Delta_n = \Delta(a_1, \dots, a_n)$ one refines Δ_n from right to left if n is odd, and from left to right if n is even. This implies that if I is any interval in $[0, 1)$ and if $\Delta_n = \Delta(a_1, \ldots, a_n)$ is the smallest cylinder containing I, then for almost all $x \in I$ either $\Delta_{n+j}(x) \subset I$, or its adjacent cylinder $\Delta_{n+j}^+ \subset I$, where $1 \leq j \leq 3$.

In general, replacing 3 in the above by $r \geq 1$, we call any piecewise defined map with properties (i), (ii) and (iii) *r-regular*. We are now ready to give a proof of Lochs' Theorem.   ∎

Proof of Theorem of Lochs. We denote the decimal map by S, its cylinder sets of order n by D_n, and $D_n(x)$ is the cylinder set of order n containing x.

For $x \in [0, 1)$, given the first n decimal digits, we find by property (iii) two T-cylinder sets $\Delta_m = \Delta_m(x)$ and Δ_{m+j}, for which

$$\Delta_{m+j} \subset D_n(x) \subset \Delta_m(x),$$

where Δ_{m+j} is either $\Delta_{m+j}(x)$, or its adjacent cylinder Δ_{m+j}^+. Then by Exercise 6.2.10,

$$\frac{-1}{n}\log 3 + \frac{1}{n}\log \lambda(\Delta_{m+r}(x)) \leq \frac{1}{n}\log \lambda(D_n(x)) \leq \frac{1}{n}\log \lambda(\Delta_m(x)).$$

Applying Lemma 6.2.2 (to both T and S) and the Theorem of Shannon-McMillan-Breiman, the result follows.   ∎

Exercise 6.2.12. Show that the above proof of Lochs' Theorem also works if we replace the continued fraction map T by any r-regular map, and the decimal transformation S by any n-ary transformation or β-transformation, or any GLS-transformation.   ∎

In fact, in 1963 Lochs showed that the first 1000 decimal digits of π yield its first 968 continued fraction digits ([Loc63]). Nowadays, with an increase in computer power, it is much easier to make such a comparison. In view of Exercise 6.2.10 we can compare several maps with maps satisfying r-regularity.

Exercise 6.2.13. Show that the *alternating Lüroth* map from Examples 2.3.8 is also r-regular, and make a computer program that gives you $m(n)$ alternating Lüroth digits if you feed it with n digits of our favorite map (the β-transformation with β equal to the golden mean). Use Corollary 6.1.13 to compare how close $m(n)/n$ gets to the ratio of the entropies. ∎

One might wonder how far Lochs' result can be extended. In a recent paper Dajani and Fieldsteel [DF] showed that Lochs' result can be extended very far indeed!

6.3 Saleski's Theorem

In Section 3.1.3 we mentioned some mixing properties, and Exercise 3.1.16 showed that Bernoullicity is the strongest possible mixing property (of the ones mentioned). In Section 3.1.3 we also mentioned a property that is close to, but weaker than, Bernoullicity: weak Bernoulli. An example of a weak Bernoulli transformation is the continued fraction map T; see also Roy Adler's paper in [PV75].

Clearly any map $Tx = nx \pmod 1$, where $n \in \mathbb{N}, n \geq 2$, is Bernoulli; see Section 1.2.2. In this section we will show that the natural extension of the β-transformation is also Bernoulli, a result due to Meir Smorodinsky [Smo73]. We will use a theorem of Alan Saleski, giving conditions under which one can conclude that a transformation is Bernoulli, given that an induced system is Bernoulli. This is an approach different from that of Smorodinsky, who showed that for each $\beta > 1$ the system

$$([0, 1), \mathcal{L}, \nu_\beta, T_\beta),$$

where $\mathcal{L}$ is the collection of Lebsegue sets of $[0,1)$, is weak Bernoulli. A deep result by N. Friedman and D. S. Ornstein [FO70] then yields that the natural extension of $([0, 1), \mathcal{L}, \nu_\beta, T_\beta)$ is a Bernoulli automorphism.

In 1973 Saleski [Sal73] obtained the following result. It refers to
a *Lebesgue space*, which is a space isomorphic to the unit interval with
Lebesgue sets and Lebesgue measure.

Theorem 6.3.1. *Let* $(X, \mathcal{F}, \mu, T)$ *be a dynamical system, with*
$(X, \mathcal{F}, \mu)$ *a non-atomic Lebesgue space. Let* $A \in \mathcal{F}$ *be a subset of*
X *of positive measure and denote by* T_A *the induced transformation of*
T *on* A. *Moreover, suppose we have that* T_A *is Bernoulli,* T *is weakly*
mixing and

$$H_{\mu_A} \left(\bigvee_{i=1}^{\infty} \bigvee_{j=1}^{\infty} T_A^i Y_j \mid \bigvee_{i=0}^{\infty} T_A^i P \right) < \infty \,,$$

where P *is a Bernoulli partition of* (A, T_A) *and*

$$Y_j = \{A - \cup_{i=1}^{j} T^{-i} A, \ A \cap \cup_{i=1}^{j} T^{-i} A\}.$$

Then T *is a Bernoulli automorphism.*

A simple corollary of Saleski's Theorem is Smorodinsky's result;
see also [DKS96], where the present approach was first given.

Corollary 6.3.2. *The completion of the system*

$$(\mathcal{H}_\beta, \mathcal{F}, \mu_\beta, \mathcal{T}_\beta),$$

as described in Section 4.5.2, is Bernoulli.

Proof. The system $(\mathcal{H}_\beta, \mathcal{F}, \mu_\beta, \mathcal{T}_\beta)$ and its completion are weakly
mixing, due to W. Parry [Par60]. Moreover, in Proposition 4.5.4 it was
shown that the induced transformation $\mathcal{W}_\beta$ is isomorphic to an appropriate GLS-transformation. Therefore $\mathcal{W}_\beta$ is Bernoulli, and in Proposi-
tion 4.5.4 we saw that $\mathcal{W}_\beta$ has Bernoulli partition $P = \mathcal{I}^\sharp := \mathcal{I} \times [0, 1]$,
where $\mathcal{I}$ is as given in Equation (4.5). The partitions Y_j of R_0 as defined

in Saleski's Theorem have the following form:

$$Y_j = \left\{ \left(0, b_0 + \frac{b_1}{\beta} + \cdots + \frac{b_j}{\beta^j} \right] \times [0, 1], \right.$$

$$\left. \left(b_0 + \frac{b_1}{\beta} + \cdots + \frac{b_j}{\beta^j}, 1 \right] \times [0, 1] \right\}.$$

Clearly, $Y_j \leq \mathcal{I}^\sharp$ for each $j \geq 1$, i.e., $\mathcal{I}^\sharp$ is a refinement of each Y_j, so that

$$\bigvee_{i=1}^{\infty} \bigvee_{j=1}^{\infty} \mathcal{W}_\beta^i Y_j \leq \bigvee_{i=0}^{\infty} \mathcal{W}_\beta^i \mathcal{I}^\sharp$$

and hence

$$H_{\mu_{R_0}} \left(\bigvee_{i=1}^{\infty} \bigvee_{j=1}^{\infty} \mathcal{W}_\beta^i Y_j \mid \bigvee_{i=0}^{\infty} \mathcal{W}_\beta^i \mathcal{I}^\sharp \right) = 0,$$

where μ_{R_0} is, by construction, Lebesgue measure on $R_0 = [0, 1)^2$; see also page 110. Thus, by Saleski's Theorem, we have that $(\mathcal{H}_\beta, \mathcal{F}, \mu_\beta, \mathcal{T}_\beta)$ is Bernoulli.   ∎

6.4   For further reading

Entropy plays a key role in ergodic theory. It is for this reason that all books on ergodic theory mentioned in Section 4.7 contain major introductions to the concept of and ideas behind the notion of entropy. We recommend two more books. One (sometimes known as the "Red Book") is *Mathematical Theory of Entropy* by N.F.G. Martin and J.W. England [ME81]. The second book is *The ergodic theory of discrete sample paths* by P.C. Shields [Shi96], where much more information can be found on weak Bernoulli transformations, and further variations on the idea of Bernoullicity, such as very weak Bernoulli.

Bibliography

[Ada79] William W. Adams, *On a relationship between the convergents of the nearest integer and regular continued fractions*, Math. Comp. **33** (1979), no. 148, 1321–1331.

[AKS81] Roy Adler, Michael Keane, and Meir Smorodinsky, *A construction of a normal number for the continued fraction transformation*, J. Number Theory **13** (1981), no. 1, 95–105.

[Apo90] Tom M. Apostol, *Modular functions and Dirichlet series in number theory*, second ed., Springer-Verlag, New York, 1990.

[AD79] J. Auslander and Y.N. Dowker, *On disjointness of dynamical systems*, Math. Proc. Cambridge Philos. Soc. **85** (1979), no. 3, 477–491.

[Bab78] K.I. Babenko, *A problem of Gauss*, Dokl. Akad. Nauk SSSR **238** (1978), no. 5, 1021–1024.

[BM66] F. Bagemihl and J. R. McLaughlin, *Generalization of some classical theorems concerning triples of consecutive convergents to simple continued fractions*, J. Reine Angew. Math. **221** (1966), 146–149.

[BJ94] Dominique Barbolosi and Hendrik Jager, *On a theorem of Legendre in the theory of continued fractions*, J. Théor. Nombres Bordeaux **6** (1994), no. 1, 81–94.

[BBDK96] Jose Barrionuevo, Robert M. Burton, Karma Dajani, and Cor Kraaikamp, *Ergodic properties of generalized Lüroth series*, Acta Arith. **74** (1996), no. 4, 311–327.

[Ber77] Anne Bertrand, *Développements en base de Pisot et répartition modulo 1*, C. R. Acad. Sci. Paris Sér. A-B **285** (1977), no. 6, A419–A421.

[BM96] Anne Bertrand-Mathis, *Nombres normaux*, J. Théor. Nombres Bordeaux **8** (1996), no. 2, 397–412.

179

[Bil65] Patrick Billingsley, *Ergodic theory and information*, John Wiley & Sons Inc., New York, 1965.

[Bil95] Patrick Billingsley, *Probability and measure*, third ed., John Wiley & Sons Inc., New York, 1995.

[Bla89] F. Blanchard, *β-expansions and symbolic dynamics*, Theoret. Comput. Sci. **65** (1989), no. 2, 131–141.

[BJW83] W. Bosma, H. Jager, and F. Wiedijk, *Some metrical observations on the approximation by continued fractions*, Nederl. Akad. Wetensch. Indag. Math. **45** (1983), no. 3, 281–299.

[BK90] Wieb Bosma and Cor Kraaikamp, *Metrical theory for optimal continued fractions*, J. Number Theory **34** (1990), no. 3, 251–270.

[BK91] Wieb Bosma and Cor Kraaikamp, *Optimal approximation by continued fractions*, J. Austral. Math. Soc. Ser. A **50** (1991), no. 3, 481–504.

[Boy96] David W. Boyd, *On the beta expansion for Salem numbers of degree 6*, Math. Comp. **65** (1996), no. 214, 861–875, S29–S31.

[Bra51] Alfred Brauer, *On algebraic equations with all but one root in the interior of the unit circle*, Math. Nachr. **4** (1951), 250–257.

[Bre89] David M. Bressoud, *Factorization and primality testing*, Springer-Verlag, New York, 1989.

[Bro76] James R. Brown, *Ergodic theory and topological dynamics*, Academic Press, New York, 1976, Pure and Applied Mathematics, No. 70.

[Cha33] D.G. Champernowne, *The construction of decimal normal in the scale of ten*, J. London Math. Soc. **8** (1933), no. 3, 254–260.

[CFS82] I. P. Cornfeld, S. V. Fomin, and Ya. G. Sinaĭ, *Ergodic theory*, Springer-Verlag, New York, 1982, Translated from the Russian by A. B. Sosinskiĭ.

[DF] K. Dajani and F. Fieldsteel, *Equipartition of interval partitions and an application to number theory*, Proc. Amer. Math. Soc., **129** (2001), no. 12, 3453–3460.

[DKS96] K. Dajani, C. Kraaikamp, and B. Solomyak, *The natural extension of the β-transformation*, Acta Math. Hungar. **73** (1996), no. 1-2, 97–109.

[DK98] Karma Dajani and Cor Kraaikamp, *A note on the approximation by continued fractions under an extra condition*, New York J. Math. **3A** (1997/98), no. Proceedings of the New York Journal of Mathematics Conference, June 9–13, 1997, 69–80 (electronic).

[DK00] K. Dajani and C. Kraaikamp, *The mother of all continued fractions*, Colloq. Math. **84/85** (2000), no. 1, 109–123.

[Dav92] H. Davenport, *The higher arithmetic.*, sixth ed., Cambridge University Press, Cambridge, 1992.

[DE52] H. Davenport and P. Erdös, *Note on normal decimals*, Canadian J. Math. **4** (1952), 58–63.

[Doe40] W. Doeblin, *Remarques sur la théorie métrique des fractions continues*, Compositio Math. **7** (1940), 353–371.

[EJK90] Pál Erdös, István Joó, and Vilmos Komornik, *Characterization of the unique expansions* $1 = \sum_{i=1}^{\infty} q^{-n_i}$ *and related problems*, Bull. Soc. Math. France **118** (1990), no. 3, 377–390.

[Fel71] William Feller, *An introduction to probability theory and its applications. Vol. II.*, second ed., John Wiley & Sons Inc., New York, 1971.

[FO70] N. A. Friedman and D. S. Ornstein, *On isomorphism of weak Bernoulli transformations*, Advances in Math. **5** (1970), 365–394.

[Fur67] Harry Furstenberg, *Disjointness in ergodic theory, minimal sets, and a problem in Diophantine approximation*, Math. Systems Theory **1** (1967), 1–49.

[Fur81] H. Furstenberg, *Recurrence in ergodic theory and combinatorial number theory*, Princeton University Press, Princeton, NJ, 1981, M. B. Porter Lectures.

[Gau76] C.F. Gauss, *Mathematisches tagebuch 1796-1814*, Akademische Verlagsgesellschaft Geest & Portig K.G., Leipzig, 1976.

[Gel59] A. O. Gelfond, *A common property of number systems*, Izv. Akad. Nauk SSSR. Ser. Mat. **23** (1959), 809–814.

[Hal50] Paul R. Halmos, *Measure Theory*, Van Nostrand Company, Inc., New York, NY, 1950.

[HW79] G. H. Hardy and E. M. Wright, *An introduction to the theory of numbers*, fifth ed., Clarendon Press, New York, 1979.

[Hla84] Edmund Hlawka, *The theory of uniform distribution*, A B Academic Publishers, Berkhamsted, 1984. With a foreword by S. K. Zaremba, Translated from the German by Henry Orde.

[Hur89] A. Hurwitz, *Über eine besondere art der kettenbruchentwicklung reeller grössen*, Acta Math. **12** (1889), 367–404.

[Ios92] Marius Iosifescu, *A very simple proof of a generalization of the Gauss-Kuzmin-Lévy theorem on continued fractions, and related questions*, Rev. Roumaine Math. Pures Appl. **37** (1992), no. 10, 901–914.

[Ios94] Marius Iosifescu, *On the Gauss-Kuzmin-Lévy theorem. I*, Rev. Roumaine Math. Pures Appl. **39** (1994), no. 2, 97–117.

[Ios95] Marius Iosifescu, *On the Gauss-Kuzmin-Lévy theorem. II*, Rev. Roumaine Math. Pures Appl. **40** (1995), no. 2, 91–105.

[Ios97] Marius Iosifescu, *On the Gauss-Kuzmin-Lévy theorem. III*, Rev. Roumaine Math. Pures Appl. **42** (1997), no. 1-2, 71–88.

[IK2002] M. Iosifescu and C. Kraaikamp, *The Ergodic Theory of Continued Fractions*. To appear, Kluwer Publishing Co., Dordrecht, the Netherlands.

[Jag82] H. Jager, *On the speed of convergence of the nearest integer continued fraction*, Math. Comp. **39** (1982), no. 160, 555–558.

[Jag86] H. Jager, *Continued fractions and ergodic theory, transcendental numbers and related topics*, RIMS Kokyuroko **599** (1986), no. 1, 55–59.

[JK89] H. Jager and C. Kraaikamp, *On the approximation by continued fractions*, Nederl. Akad. Wetensch. Indag. Math. **51** (1989), no. 3, 289–307.

[JL88] Hendrik Jager and Pierre Liardet, *Distributions arithmétiques des dénominateurs de convergents de fractions continues*, Nederl. Akad. Wetensch. Indag. Math. **50** (1988), no. 2, 181–197.

[Kak43] Shizuo Kakutani, *Induced measure preserving transformations*, Proc. Imp. Acad. Tokyo **19** (1943), 635–641.

[KKK90] Sofia Kalpazidou, Arnold Knopfmacher, and John Knopfmacher, *Lüroth-type alternating series representations for real numbers*, Acta Arith. **55** (1990), no. 4, 311–322.

[KKK91] Sofia Kalpazidou, Arnold Knopfmacher, and John Knopfmacher, *Metric properties of alternating Lüroth series*, Portugal. Math. **48** (1991), no. 3, 319–325.

[Kam82] Teturo Kamae, *A simple proof of the ergodic theorem using nonstandard analysis*, Israel J. Math. **42** (1982), no. 4, 284–290.

[KW82] Yitzhak Katznelson and Benjamin Weiss, *A simple proof of some ergodic theorems*, Israel J. Math. **42** (1982), no. 4, 291–296.

[Kel98] Gerhard Keller, *Equilibrium states in ergodic theory*, Cambridge University Press, Cambridge, 1998.

[Khi35] A. Ya. Khintchine, *Metrische kettenbruchproblemen*, Compositio Math. **1** (1935), 361–382.

[Khi63] A. Ya. Khintchine, *Continued fractions*, P. Noordhoff Ltd., Groningen, 1963. Translated by Peter Wynn.

[KT66] J. F. C. Kingman and S. J. Taylor, *Introduction to measure and probability*, Cambridge University Press, London, 1966.

[Kit98] Bruce P. Kitchens, *Symbolic dynamics*, Springer-Verlag, Berlin, 1998.

[Kol58] A. N. Kolmogorov, *A new metric invariant of transient dynamical systems and automorphisms in Lebesgue spaces*, Dokl. Akad. Nauk SSSR (N.S.) **119** (1958), 861–864.

[Kol59] A. N. Kolmogorov, *Entropy per unit time as a metric invariant of automorphisms*, Dokl. Akad. Nauk SSSR **124** (1959), 754–755.

[Kra90] Cor Kraaikamp, *On the approximation by continued fractions. II*, Indag. Math. (N.S.) **1** (1990), no. 1, 63–75.

[Kra91] Cor Kraaikamp, *A new class of continued fraction expansions*, Acta Arith. **57** (1991), no. 1, 1–39.

[Kra93] Cornelis Kraaikamp, *Maximal S-expansions are Bernoulli shifts*, Bull. Soc. Math. France **121** (1993), no. 1, 117–131.

[Kre85] Ulrich Krengel, *Ergodic theorems*, Walter de Gruyter & Co., Berlin, 1985. With a supplement by Antoine Brunel.

[KN74] L. Kuipers and H. Niederreiter, *Uniform distribution of sequences*, Wiley-Interscience, New York, 1974.

[Kus28] R.O. Kusmin, *Sur un probléme de Gauss*, Atti Congr. Bologne **6** (1928), 83–89.

[L29] P. Lévy, *Sur le loi de probabilité dont dependent les quotients complets et incomplets d'une fraction continue*, Bull. Soc. Math. de France **57** (1929), 178–194.

[LM95] Douglas Lind and Brian Marcus, *An introduction to symbolic dynamics and coding*, Cambridge University Press, Cambridge, 1995.

[Loc63] Gustav Lochs, *Die ersten 968 Kettenbruchnenner von π*, Monatsh. Math. **67** (1963), 311–316.

[Loc64] Gustav Lochs, *Vergleich der Genauigkeit von Dezimalbruch und Kettenbruch*, Abh. Math. Sem. Univ. Hamburg **27** (1964), 142–144.

[Lür83] J. Lüroth, *Ueber eine eindeutige entwickelung von zahlen in eine unendliche reihe*, Math. Annalen **21** (1883), 411–423.

[Mn87] Ricardo Mañé, *Ergodic theory and differentiable dynamics*, Springer-Verlag, Berlin, 1987. Translated from Portuguese by Silvio Levy.

[ME81] Nathaniel F. G. Martin and James W. England, *Mathematical theory of entropy*, Addison-Wesley, Reading MA, 1981.

[Min73] B. Minnigerode, *Über eine neue methode, die pellsche gleichung aufzulösen*, Nachr. Göttingen (1873).

[Moe82] R. Moeckel, *Geodesics on modular surfaces and continued fractions*, Ergodic Theory Dynamical Systems **2** (1982), no. 1, 69–83.

[Nak81] Hitoshi Nakada, *Metrical theory for a class of continued fraction transformations and their natural extensions*, Tokyo J. Math. **4** (1981), no. 2, 399–426.

[Nak95] Hitoshi Nakada, *Continued fractions, geodesic flows and Ford circles*, (Okayama/Kyoto, 1992); Plenum, New York, 1995, pp. 179–191.

[NIT77] Hitoshi Nakada, Shunji Ito, and Shigeru Tanaka, *On the invariant measure for the transformations associated with some real continued-fractions*, Keio Engrg. Rep. **30** (1977), no. 13, 159–175.

[Nol90] V. N. Nolte, *Some probabilistic results on the convergents of continued fractions*, Indag. Math. (N.S.) **1** (1990), no. 3, 381–389.

[Par60] W. Parry, *On the β-expansions of real numbers*, Acta Math. Acad. Sci. Hungar. **11** (1960), 401–416.

[Pet89] Karl Petersen, *Ergodic theory*, Cambridge University Press, Cambridge, 1989, Corrected reprint of the 1983 original.

[PV75] E. Phillips and S. Varadhan (eds.), *Ergodic theory*, Courant Institute of Mathematical Sciences, New York University, New York, 1975. A seminar held at the Courant Institute 1973–1974. With contributions by S. Varadhan, E. Phillips, S. Alpern, N. Bitzenhofer and R. Adler.

[PY98] Mark Pollicott and Michiko Yuri, *Dynamical systems and ergodic theory*, Cambridge University Press, Cambridge, 1998.

[PS64] G. Pólya and G. Szegö, *Aufgaben und Lehrsätze aus der Analysis. Band II: Funktionentheorie. Nullstellen. Polynome. Determinanten. Zahlentheorie*, Springer-Verlag, Berlin, 1964, Dritte berichtigte Auflage. Die Grundlehren der Mathematischen Wissenschaften, Band 20.

[Pos60] A. G. Postnikov, *Arithmetic modeling of random processes*, Trudy Mat. Inst. Steklov. **57** (1960), 84.

[Rad83] Hans Rademacher, *Higher mathematics from an elementary point of view*, Birkhäuser, Boston, 1983. Edited by D. Goldfeld, with notes by G. Crane.

[Rén57] A. Rényi, *Representations for real numbers and their ergodic properties*, Acta Math. Acad. Sci. Hungar. **8** (1957), 477–493.

[RS92] Andrew M. Rockett and Peter Szüsz, *Continued fractions*, World Scientific, River Edge, NJ, 1992.

[Roh61] V. A. Rohlin, *Exact endomorphisms of a Lebesgue space*, Izv. Akad. Nauk SSSR Ser. Mat. **25** (1961), 499–530.

[Roy88] H. L. Royden, *Real analysis*, third ed., Macmillan, New York, 1988.

[Rud87] Walter Rudin, *Real and complex analysis*, third ed., McGraw-Hill, New York, 1987.

[Rud90] Daniel J. Rudolph, *Fundamentals of measurable dynamics*, Clarendon Press, New York, 1990.

[RN51] C. Ryll-Nardzewski, *On the ergodic theorems. II. Ergodic theory of continued fractions*, Studia Math. **12** (1951), 74–79.

[Sal63] Raphaël Salem, *Algebraic numbers and Fourier analysis*, D. C. Heath, Boston MA, 1963.

[Sal73] Alan Saleski, *On induced transformations of Bernoulli shifts*, Math. Systems Theory **7** (1973), 83–96.

[Sch80a] Klaus Schmidt, *On periodic expansions of Pisot numbers and Salem numbers*, Bull. London Math. Soc. **12** (1980), no. 4, 269–278.

[Sch80b] Wolfgang M. Schmidt, *Diophantine approximation*, Springer, Berlin, 1980.

[Sch68] Fritz Schweiger, *Metrische Theorie einer Klasse zahlentheoretischer Transformationen.*, Acta Arith. **15** (1968), 1–18.

[Sch70] Fritz Schweiger, *Metrische Theorie einer Klasse zahlentheoretischer Transformationen: Corrigendum*, Acta Arith. **16** (1969/1970), 217–219.

[Sch95] Fritz Schweiger, *Ergodic theory of fibred systems and metric number theory*, Clarendon Press, New York, 1995.

[Seg45] B. Segre, *Lattice points in infinite domains and asymmetric Diophantine approximations*, Duke Math. J. **12** (1945), 337–365.

[Ser82] Caroline Series, *Non-Euclidean geometry, continued fractions, and ergodic theory*, Math. Intelligencer **4** (1982), no. 1, 24–31.

[Ser85] Caroline Series, *The geometry of Markoff numbers*, Math. Intelligencer **7** (1985), no. 3, 20–29.

[Sha48] C. E. Shannon, *A mathematical theory of communication*, Bell System Tech. J. **27** (1948), 379–423, 623–656.

[Shi96] Paul C. Shields, *The ergodic theory of discrete sample paths*, American Mathematical Society, Providence, RI, 1996.

[Shi71] Goro Shimura, *Introduction to the arithmetic theory of automorphic functions*, Publications of the Mathematical Society of Japan, No. 11. Iwanami Shoten, Tokyo, 1971, Kanô Memorial Lectures, No. 1.

[Smo73] M. Smorodinsky, *β-automorphisms are Bernoulli shifts*, Acta Math. Acad. Sci. Hungar. **24** (1973), 273–278.

[Szü61] P. Szüsz, *Über einen Kusminschen Satz*, Acta Math. Acad. Sci. Hungar. **12** (1961), 447–453.

[SV94] Peter Szüsz and Bodo Volkmann, *A combinatorial method for constructing normal numbers*, Forum Math. **6** (1994), no. 4, 399–414.

[Ton83] Jingcheng Tong, *The conjugate property of the Borel theorem on Diophantine approximation*, Math. Z. **184** (1983), no. 2, 151–153.

[Vit95] Paul Vitányi, *Randomness*, CWI Quarterly **8** (1995), no. 1, 67–82.

[Wag95] Gerold Wagner, *On rings of numbers which are normal to one base but non-normal to another*, J. Number Theory **54** (1995), no. 2, 211–231.

[Wal82] Peter Walters, *An introduction to ergodic theory*, Springer-Verlag, New York, 1982.

[Wir74] Eduard Wirsing, *On the theorem of Gauss-Kusmin-Lévy and a Frobenius-type theorem for function spaces*, Acta Arith. **24** (1973/74), 507–528. Collection of articles dedicated to Carl Ludwig Siegel on the occasion of his seventy-fifth birthday, V.

Index